The MRCP PACES Handbook

SAIRA GHAFUR
SpR Respiratory Medicine

RICHARD KITCHEN
SpR Clinical Oncology

PARMINDER JUDGE
SpR Nephrology

and

SAMUEL BLOWS
SpR Medicine for the Elderly

Foreword by

BRYAN WILLIAMS
Professor of Medicine
University of Leicester and University Hospitals of Leicester NHS Trust

Radcliffe Publishing
London • New York

Radcliffe Publishing Ltd
33–41 Dallington Street
London
EC1V 0BB
United Kingdom

www.radcliffepublishing.com
Electronic catalogue and worldwide online ordering facility.

British Library Cataloguing in Publication Data

A catalogue record for this book is available from the British Library.

ISBN-13: 978 184619 497 9

The paper used for the text pages of this book is FSC certified. FSC® (The Forest Stewardship Council®) is an international network to promote responsible management of the world's forests.

Typeset by Pindar NZ, Auckland, New Zealand
Printed and bound by TJI Digital, Padstow, Cornwall, UK

Contents

Foreword

My own MRCP examinations (Part 2 in those days) were 25 years ago, or 'a quarter of a century', as my daughter is likely to remind me! Despite the passage of time and an eventful career along the way, remarkably, I can still recall the individual cases that I examined and presented at my MRCP clinical exam. I am sure this is the same for many of my colleagues and this perhaps underscores what a 'life event' the whole process can be. Not to say that we did not have a good time along the way, working for the examinations. I recall fondly the group of eight of us who were all SHOs in Leicester, getting together for evening teaching sessions after the day's work was done, enthusiastically organised by the Registrars or Consultants.

Things have changed and, of course, the examination process has moved on but the basic reason for the examination remains. It is designed to test knowledge, clinical skills, and the ability to recognise disease patterns, establish differential diagnoses, and define appropriate investigation and treatment. Moreover, the ability to do this rapidly whilst thinking on your feet – precisely the skills that defined a good medical registrar then, and still does today. So, the format of the test has changed but the hurdle is set at the same height. I recall reading many handbooks to help me prepare and many were good. However, none were written in such a way that they were directly orientated towards the MRCP clinical examination. This is where *The MRCP PACES Handbook* comes into its own. The authors have identified common clinical conditions in different disease areas – cases frequently encountered in clinical examinations that lend themselves to test clinical acumen and competence. The style is succinct, not minimalist but sufficient. The presentation is fresh and logical. I think the authors have done an outstanding job in introducing the setting for the PACES examination and the expectations of the examiners. We did not have such a concise, focused and well-presented book of this genre in my day. Had it been available, *The MRCP PACES Handbook* would certainly have been very popular then, as I am sure it will be today. Of course, alone, this book will not secure your success, even the brightest need to do the work, but this book will reduce the risk of failure due to poor preparation and lack of insight into what is required to succeed. The authors, I know, have been motivated to make the challenge of the PACES examination less daunting to those that follow them, than it appeared to them. I think they have achieved that objective and I wish them every success with this gem of a book; they deserve it.

Bryan Williams MD FRCP FAHA FESC
Professor of Medicine
University of Leicester and
University Hospitals of Leicester
NHS Trust
January 2011

Preface

PACES should be seen as the defining moment of your medical training when you have been accepted into the fold – the final hurdle you need to jump over before gaining the elusive title of 'Medical Registrar'.

This book is designed to be a comprehensive study guide that will be invaluable to candidates studying for PACES. They will be able to use this book at the patient's bedside to guide clinical examination as well as having enough detail to cover pertinent points in the given case.

Each of the chapters contains hints and tips on how to tackle each of the examinations and a summary at the end to consolidate key learning points. The cases are those which are most frequently encountered in the examination and are all set out in a standardised format. Each chapter has been reviewed by Consultants within the specialty, many of whom are Royal College examiners. The information is up to date and is supported by evidence-based literature and recent guidelines.

October 2009, has seen the introduction of a new marking scheme for the exam as well as a new format for Station 5. The Station 5 cases termed 'Integrated clinical assessment' encompass all key areas in which the candidates are expected to perform to gain maximum marks. *The MRCP PACES Handbook* is one of the few books currently available highlighting this fundamental change to the exam.

Passing PACES not only requires an in-depth knowledge of the subject but also requires the fine art of being able to present the case in a concise and coherent fashion. Each case contains a section on presentation which will aid the candidate in perfecting this skill.

PACES is feared as the unachievable milestone; the key to passing the exam is hard work and sheer dedication! The blood, sweat and tears will all be worth it in the end when you have attained the highly sought title of MRCP (UK).

If you follow in the footsteps of those who have gone before you, you are guaranteed to succeed!

Good luck . . . go forth and conquer!

SG, RK, PJ and SB
January 2011

About the authors

Dr Saira Ghafur graduated from the University of Dundee Medical School. She is a Specialist Registrar in Respiratory Medicine on the South Yorkshire and the Humber rotation. She has an interest in pulmonary hypertension.

Dr Richard Kitchen graduated from the University of Leicester Medical School. He is a Specialist Registrar in Clinical Oncology on the West Midlands rotation. He has a particular interest in the palliative aspects of care.

Dr Parminder Judge graduated from the University of Birmingham Medical School. She is a Nephrology Registrar on the Oxford rotation. She has a particular interest in vasculitis and acute kidney injury.

Dr Samuel Blows graduated from the University of Leicester Medical School. He is a Specialist Registrar in Medicine for the Elderly on the North West London rotation. He has a specialist interest in dementia.

The authors all worked together at the University Hospitals of Leicester as junior doctors, during which time they all passed their PACES exam and developed their interest in medical education.

Acknowledgements

We would like to thank the following people for all their help, support, specialty knowledge and guidance in making this book possible:

Professor Bryan Williams
Professor of Medicine, University Hospitals of Leicester

Dr Simon Conroy
Consultant Geriatrician, Senior Lecturer, University Hospitals of Leicester

Dr Mark Ardron
Consultant Geriatrician, University Hospitals of Leicester

Professor Mike Morgan
Chairman of the British Thoracic Society, Professor of Respiratory Medicine, University Hospitals of Leicester

Dr David Sharman
Specialist Registrar in Cardiology, Oxford Radcliffe Hospitals NHS Trust

Dr Peter Topham
Consultant Nephrologist, Clinical Senior Lecturer, University Hospitals of Leicester

Dr Jessica Williams
Consultant Gastroenterologist, Derby Royal Infirmary

Dr David Sprigings
Consultant Cardiologist, Northampton General Hospital

Dr Sherrini Ahmed
Specialist Registrar in Neurology

Dr Iqbal Khan
Consultant Gastroenterologist, Northampton General Hospital

Dr Claire Chapman
Consultant Haematologist, University Hospitals of Leicester

Dr Ajay M Verma
Specialist Registrar in Gastroenterology, University Hospitals of Leicester

Dr Patricia Hooper
Specialist Registrar in Gastroenterology, University Hospitals of Leicester

The Medical Illustration Department
University Hospitals of Leicester

And the patients who kindly allowed us to use their images in this book.

List of abbreviations

A2	aortic component of heart sound two
AA	amyloid A
ABG	arterial blood gas
ABPA	allergic broncho-pulmonary aspergillosis
ACE	angiotensin converting enzyme
ACTH	adrenocorticotrophic hormone
ADH	antidiuretic hormone
ADL	activities of daily living
AF	atrial fibrillation
AFP	alpha-fetoprotein
AKI	acute kidney injury
ALD	adrenoleukodystrophy
ALP	alkaline phosphatase
ALS	amyotrophic lateral sclerosis
AMTS	abbreviated mental test score
APKD	adult polycystic kidney disease
AR	aortic regurgitation
ARDS	acute respiratory distress syndrome
AS	aortic stenosis
ASA	aminosalicylic acid
ASD	atrial septal defect
AV	arteriovenous
AVR	aortic valve replacement
AXR	abdominal X-ray
BCC	basal cell carcinoma
BP	blood pressure
BTS	British Thoracic Society
CABG	coronary artery bypass graft
CCF	congestive cardiac failure
CF	cystic fibrosis
Ch	chromosome
CLL	chronic lymphocytic leukaemia
CML	chronic myeloid leukaemia
CMV	cytomegalovirus
CN	cranial nerve

CO	carbon monoxide
COMT	catechol-o-methyl transferase
COPD	chronic obstructive pulmonary disease
CRF	chronic renal failure
CRP	C-reactive protein
CSF	cerebrospinal fluid
CT	computerised tomography
CVA	cerebrovascular accident
CXR	chest X-ray
DEXA	dual-energy X-ray absorptiometry
DMARD	disease-modifying antirheumatoid drug
DVLA	Driver and Vehicle Licensing Agency
EBUS	endobronchial ultrasound
EBV	Epstein–Barr's virus
ECG	electrocardiogram
EDM	end-diastolic murmur
EMG	electromyogram
ESR	erythrocyte sedimentation rate
ESRF	end-stage renal failure
FBC	full blood count
FEV_1	forced expiratory volume in 1st second
FiO_2	fraction of inspired oxygen
FSH	follicle stimulating hormone
FVC	forced vital capacity
GBM	glomerular basement membrane
GBS	Guillian–Barré syndrome
GGT	gamma glutamyl transpeptidase
GH	growth hormone
GIM	general internal medicine
GNRH	gonadotrophin releasing hormone
GP	general practitioner
HCM	hypertrophic cardiomyopathy
HIV	human immunodeficiency virus
HPOA	hypertrophic pulmonary osteoarthropathy
IBS	irritable bowel syndrome
ICU	intensive care unit
IE	infective endocarditis
IGF-1	insulin-like growth factor 1
IIH	idiopathic intracranial hypertension
IL	interleukin

INO	internuclear ophthalmoplegia
INR	international normalised ratio
ITP	idiopathic thrombocytopenic purpura
IVDU	intravenous drug user
IV	intravenous
JVP	jugular venous pressure
KCO	carbon monoxide transfer coefficient
LA	left atrium
LBBB	left bundle branch block
LDH	lactate dehydrogenase
LFTs	liver function tests
LH	luteinising hormone
LMN	lower motor neurone
LMWH	low molecular weight heparin
LP	lumbar puncture
LTOT	long-term oxygen therapy
LUQ	left upper quadrant
LV	left ventricle
LVF	left ventricular failure
MCPJ	metacarpophalangeal joint
MCV	mean cell volume
MDT	multidisciplinary team
MHC	major histocompatibility antigen
MMF	mycophenolate mofetil
MND	motor neurone disease
MR	mitral regurgitation
MRI	magnetic resonance imaging
MRC	Medical Research Council
MS	multiple sclerosis
MVP	mitral valve prolapse
NCS	nerve conduction study
NICE	National Institute for Health and Clinical Excellence
NIH	National Institute of Health
NIPPV	non-invasive positive pressure ventilation
NSCLC	non-small-cell lung cancer
NSTEMI	non-ST elevation myocardial infarction
NYHA	New York Heart Association
OCP	oral contraceptive pill
OD	once daily
OGD	oesophagogastroduodenoscopy

ON	once nightly
OSA	obstructive sleep apnoea
OT	occupational therapist
P2	pulmonary component of heart sound two
PBC	primary biliary cirrhosis
PD	peritoneal dialysis
PDA	patent ductus arteriosus
PEG	percutaneous endoscopic gastrostomy
PET	positron emission tomography
PFTs	pulmonary function tests
PLS	primary lateral sclerosis
PMA	progressive muscular atrophy
PPI	proton pump inhibitor
PPM	permanent pacemaker
PR	pulmonary regurgitation
PRN	as needed (*pro re nata*)
PTH	parathyroid hormone
QOL	quality of life
RA	rheumatoid arthritis
RIF	right iliac fossa
S1	heart sound one
S2	heart sound two
S3	heart sound three
S4	heart sound four
SAH	subarachnoid haemorrhage
SALT	speech and language therapist
SCC	squamous cell carcinoma
SCL-70	anti-topoisomerase antibodies
SCLC	small-cell lung cancer
SIADH	syndrome of inappropriate antidiuretic hormone secretion
SIGN	Scottish Intercollegiate Guidelines Network
SLE	systemic lupus erythematosus
SOL	space-occupying lesion
SVCO	superior vena caval obstruction
TAVI	transcutaneous aortic valve implantation
TB	tuberculosis
TFTs	thyroid function tests
TIA	transient ischaemic attack
TLC	total lung capacity

TLCO	transfer factor of the lung for carbon monoxide
TNF	tumour necrosis factor
TOE	transoesophageal echocardiogram
TPO	thyroid peroxidase
TR	tricuspid regurgitation
TRH	thyroid releasing hormone
TSH	thyroid stimulating hormone
tTG	tissue transglutaminase
U1-RNP	U1-ribonucleoprotein
U&Es	urea and electrolytes
UC	ulcerative colitis
UMN	upper motor neurone
US	ultrasound
USS	ultrasound scan
VATS	video-assisted thoracic surgery
VP	ventriculoperitoneal
VSD	ventricular septal defect
WHO	World Health Organization

The PACES examination

The PACES (Practical Assessment of Clinical Examination Skills) examination consists of five 20-minute stations that make up a carousel. The examination lasts for a total of 125 minutes which includes a five-minute break between each of the stations.

Stations 1 (respiratory/abdominal) and 3 (cardiovascular/neurology) are clinical scenarios lasting 10 minutes. A maximum of six minutes is allowed for examination, followed by four minutes of discussion. Candidates will be given written instructions prior to examining the patient.

Stations 2 (history-taking skills) and 4 (communication skills and ethics) last 20 minutes. In these stations 14 minutes will be allowed for history taking or communicating with the patient. One minute of reflection is allowed prior to five minutes' discussion with the examiners.

Station 5 (integrated clinical assessment) changed in 2009 to two brief clinical scenarios of 10 minutes each. Eight minutes are allocated for focused history taking, examination and responding to the patient's concerns. The remaining two minutes are for discussing examination findings, diagnosis, and patient's concerns with the examiners. Candidates can be assessed on a wide range of clinical problems which are based on the GIM curriculum. A high proportion of the overall marks have been allotted to this station.

There will be two examiners present at each station. Candidates will be assessed in seven key areas. They are not all assessed at every station. The key areas include:

- physical examination
- identifying physical signs
- clinical communication
- differential diagnosis
- clinical judgement
- managing patient concerns
- managing patient welfare.

Examiners will mark candidates as satisfactory, borderline or unsatisfactory in each area. It is necessary to reach a minimum standard in each of these areas, as well as meet the overall pass mark.

Further information can be sought from the MRCP UK website (www.mrcpuk.org). The website provides details of examination dates and fees and also gives a comprehensive outline of what the exam will involve. There are copies of the marksheets available for each station. The website also gives example scenarios for each station.

Tips for candidates

- Dress sense is important; you want to stand out because of your clinical skills and not the way you dress! Men should arrive in a suit with polished shoes and a sensible tie. Women should also wear a smart suit and ensure that the hem line is kept at a conservative length. At the same time you must be comfortable. Examination centres take infection control measures very seriously, so candidates may be asked to remove ties, watches and also to roll up sleeves.
- Give yourself plenty of time to get to the examination centre and make sure your equipment is packed the night before.
- If taking the exam far from home, book a hotel close to the examination centre for the night before. You do not want to be worrying about train delays or traffic on the day of the exam.
- Act formally in the examination; it is not a place for jokes (with the examiners or patients).
- Always be polite and courteous to the patients and thank them when you are finished.
- It doesn't matter which order you proceed through the stations, everyone has to do all five stations!
- You must try to forget about any bad station you have had, as doing exceptionally well at another one can make up for it.
- Bring with you to the exam:
  - — your entry card
  - — your stethoscope
  - — your own ophthalmoscope (if you have one)
  - — a pen
  - — a watch.
- All other equipment will be provided.

Station 1: Respiratory

Contents

Hints for the respiratory station

- If you are going to comment on respiratory rate, make sure you count it.
- Be aware of inhaler colours and design, and know what they contain.
- Look for clues around the bedside, e.g. spacer device, peakflow meter.
- Always look in the sputum pot if present.
- If the patient is on oxygen, comment on delivery method and the percentage inspired.
- If commenting on breathlessness, try to use the MRC scale to grade it.
- Examine the patient starting from the back, as findings are more likely to be picked up from the back (important, if time is running out).
- Try to look for underlying causes of the pathology identified.
- Look for signs of cor pulmonale where appropriate.
- Complete the examination by telling the examiner you would like to check oxygen saturations, peak flow and spirometry results where appropriate.

Bronchiectasis

'Please examine this patient who has a cough.'

Findings

- General
  - age (may help with underlying diagnosis), cachexia, dyspnoea, inhalers and sputum pot
- Peripheral
  - clubbing, signs of cor pulmonale
- Chest
  - coarse inspiratory crackles (which may clear with coughing), wheeze, situs inversus

Presentation

'The primary diagnosis in this patient is bronchiectasis. She is producing purulent sputum. She has clubbing, and has coarse inspiratory crepitations in both lower zones. I cannot elicit an obvious cause, but I would ask about childhood infections.'

- Comment on the patient's productive cough.
- Look out for a sputum pot and be sure to look at the contents.
- Look for an obvious underlying cause including cystic fibrosis and COPD.
- If there is no obvious underlying cause, mention that the top three causes are postinfective, cystic fibrosis and COPD.

Investigations

- Diagnosis
  - chest X-ray: may be normal; dilated bronchi with thickened walls ('tramline' shadowing); ring shadowing
  - high-resolution CT chest: bronchial dilatation with wall thickening ('signet ring' appearance)
  - sputum: culture and sensitivities; *Pseudomonas aeruginosa* is important
  - spirometry: may show an obstructive picture

- Cause
  - investigations for cystic fibrosis (2 measurements of sweat chloride and CFTR genetic mutation analysis)
  - immunoglobulin screen
  - ABPA screen
  - saccharin test for mucociliary clearance
  - CT of the chest will locate obstructive lesions and fibrosis

Management

- Specialist physiotherapy with a daily routine for patients
- Antibiotics
  - treatment of acute exacerbations often requires intravenous therapy with antipseudomonal agents (piperacillin, ceftazidime, carbapenems, aminoglycosides)
  - oral ciprofloxacin may be used
  - long-term antibiotics may include nebulised colistin and oral macrolides
- Bronchodilators
  - in those with airflow limitation
- Mucolytics
  - carbocisteine
- Surgery
  - bronchiectasis is rarely sufficiently localised to be amenable to surgery but may be an option in a very small minority of cases

Questions

1 'What is bronchiectasis?'
  - Abnormal and permanently dilated airways with bronchial wall thickening
  - This is manifest as a cough with copious purulent sputum

2 'What are the causes of bronchiectasis?'
  - Postinfective bronchial damage: bacterial and viral pneumonias, including measles and pertussis
  - Mucociliary clearance defects: cystic fibrosis, primary ciliary dyskinesia/Kartagener's syndrome, Young's syndrome
  - Immunodeficiency: primary (immunoglobulin deficiency) and secondary (HIV infection, malignancy)
  - Inflammatory (granuloma and fibrosis): idiopathic pulmonary fibrosis, tuberculosis, sarcoid

— Mechanical: obstruction (tumour, foreign body), COPD, traction bronchiectasis
— Immunological response: allergic bronchopulmonary aspergillosis

3 'What is the differential diagnosis of bilateral lower-zone crepitations?'
— Bronchiectasis: coarse crackles heard in early–mid inspiration
— Lung fibrosis: fine, late crackles; look for clubbing, dry cough and cyanosis, as well as a cause
— Pulmonary oedema: fine/coarse crackles; look for evidence of cardiac disease
— Bilateral pneumonia: coarse crackles; look for pyrexia and bronchial breathing

Key points

- Clubbing, along with coarse inspiratory crackles (that clear with coughing), are the hallmark clinical features of bronchiectasis.
- 'Signet ring' appearance of dilated bronchi on a CT scan of the thorax is diagnostic.
- Ensure that you know the common causes of bronchiectasis.

Chronic obstructive pulmonary disease

'Please examine this breathless patient.'

Findings

- General
  - dyspnoea, respiratory distress, prolonged expiratory time, wheeze, presence of inhalers and oxygen
- Peripheral
  - cyanosis, tar staining, bronchodilator tremor, flap
- Chest
  - hyperexpanded chest, reduced cricosternal distance, hyperresonance and reduced breath sounds (over bullae), wheeze, Hoover's sign (inward movement of the lower rib-cage during inspiration-associated with COPD)

Presentation

'This gentleman is dyspnoeic at rest and is on oxygen therapy. He has nicotine-stained fingertips. There is central cyanosis with widespread wheeze throughout both lungs. I note the presence of inhalers at the bedside. The diagnosis is COPD.'

- Comment on external clues including inhalers, oxygen, nebulisers and a sputum pot.
- Comment on any features of respiratory distress and ensure the respiratory rate is counted.
- Look for features of cor pulmonale (raised JVP, loud P2, peripheral oedema), polycythaemia (plethora), infection (consolidation) and Cushing's syndrome (corticosteroid use).

Investigations

- Diagnosis
  - spirometry with bronchodilator response, arterial blood gas (type 1 or 2 respiratory failure is possible), CXR (hyperinflation, flat hemidiaphragms, bullae, large pulmonary arteries)

- Complications
  - echocardiogram (pulmonary hypertension), CXR (infection/pneumothorax)

Management

- Acute
  - Oxygen therapy (with caution), nebulised bronchodilators, corticosteroids, antibiotics for infection, theophylline, physiotherapy, consideration of non-invasive positive-pressure ventilation (NIPPV) and intubation
- Chronic: depends on the severity of airway obstruction
  - mild
    - PRN inhalers (salbutamol and/or ipratropium)
  - moderate
    - long-acting inhaled bronchodilator (tiotropium and/or β2-agonist)
    - consider inhaled steroids
    - consider oral aminophylline
  - severe
    - consider home nebulisers
- Extras
  - stop smoking, nutritional management, pulmonary rehabilitation,[1] vaccinations, home oxygen/long-term oxygen therapy (LTOT), surgery, social support, note MRC score for breathlessness

Questions

1 'What is cor pulmonale? What is its significance?'
  - Right-sided cardiac dysfunction secondary to pulmonary hypertension. The pulmonary hypertension must be of a respiratory cause (chronic lung disease, pulmonary vasculature disorders, neuromuscular disease affecting the respiratory system)
  - Untreated cor pulmonale causes right-sided heart failure and death

2 'What are the indications for long-term oxygen therapy (LTOT)?'
  - LTOT describes oxygen given for >16 hours/day, with the aim of achieving a P_ao_2 >8 kPa

- It is indicated for those patients with a $P_{a}O_2$ <7.3 kPa on two consecutive readings at least three weeks apart (in a stable patient), or for a $P_{a}O_2$ 7.3–8 kPa in a patient with cor pulmonale
- It has been shown to improve mortality

3 'What would full pulmonary function tests show (including severity)?'
- Spirometry: obstructive picture (decreased FEV_1:FVC ratio)
- Total lung capacity: may be increased
- Residual volume: may be increased
- Gas transfer coefficient: may be reduced
- Flow volume loop: intrathoracic obstruction ('scalloping' of curve)

Key points

- Be able to identify what medications inhalers contain, from their colour/design.
- Ensure that you know the criteria for LTOT.

Consolidation

'Please examine this patient's respiratory system.'

Findings

- General
  - oxygen, nebulisers, sputum pot, haemoptysis, herpes labialis
- Peripheral
  - tachypnoea, tachycardia, pyrexia
- Chest
  - decreased chest expansion, dullness to percussion over affected lobe, bronchial breathing, crackles, increased vocal resonance, whispering pectoriloquy

Presentation

'This patient has consolidation at his right base. He is receiving oxygen therapy and has evidence of respiratory distress. There is dullness to percussion at the right lung base with decreased breath sounds, bronchial breathing and coarse crepitations.'

- Always remember that if the information about the case is handwritten, it is likely that a patient has withdrawn from the exam at short notice and a last-minute replacement has been recruited from the wards!
- Such a patient is likely to be more unwell, therefore comment on any features of respiratory distress, note any infusions or nebulisers and check the FiO_2 being delivered.
- The commonest cause of consolidation is pneumonia.

Investigations

- Diagnosis
  - CXR, ABG, sputum culture, routine bloods, blood cultures, atypical pneumonia screen (urine for legionella and pneumococcal antigens, mycoplasma serology) if indicated

Management

- Use the CURB 65 score to assess the severity of the pneumonia
  - confusion (AMTS ≤ 8)

- urea >7 mmol/L
- respiratory rate ≥30 breaths/minute
- blood pressure (systolic <90 mmHg or diastolic <60 mmHg)
- age >65 years

- A score of 0 or 1: treat as an outpatient
- A score of 2: possible short stay in hospital
- A score of 3–5: requires hospitalisation and may require critical care intervention
- Treatment is with antimicrobials and oxygen therapy where required
- Antimicrobial therapy is guided by local trust guidelines, but generally use amoxicillin +/– a macrolide (to cover atypical organisms)
- Patients should be followed up after a 6 week interval with a repeat CXR

Questions

1 'What are the common organisms causing community-acquired pneumonia?'
   - Common: *Streptococcus pneumoniae, Haemophilus influenzae, Staphylococcus aureus*
   - Atypical: *Mycoplasma pneumoniae, Legionella pneumophila, Chlamydophila pneumoniae, Chlamydophila psittaci, Coxiella burnetii*
   - Viruses: influenza, CMV and varicella-zoster

2 'List possible complications of pneumonia.'
   - Para-pneumonic effusion, empyema, cavitation, lung abscess, septic shock, respiratory failure/ARDS, hepatitis, haemolytic anaemia, erythema multiforme

3 'What is the difference between an empyema and a complicated para-pneumonic effusion?'
   - An empyema is pus in the pleural cavity with a pH <7.2
   - A complicated para-pneumonic effusion has a pH <7.2 but is clear

4 'How would a consolidation be differentiated from an effusion on examination?'
   - Tactile vocal fremitus: sound transmission is increased through tissue (consolidation) and decreased through fluid (pleural effusion)
   - Whispering pectoriloquy: is indicative of consolidation when whispered sounds are heard clearly through affected lung tissue

Key points

- Look closely at the information presented to the candidate, i.e. is it handwritten.
- Patients from the ward may be more unwell; comment on any respiratory distress.
- Be aware of complications associated with pneumonia.

Cystic fibrosis

'This patient has presented with repeated chest infections. Please examine their respiratory system.'

Findings

- General
  - young patient, short stature, cachexia, pallor, dyspnoea, permanent vascular access (peripheral/central), presence of inhalers and sputum pot, pinpricks from blood glucose measurement
- Peripheral
  - gross finger clubbing, signs of cor pulmonale
- Chest
  - lung transplant scar, coarse inspiratory crackles (which may clear with coughing), wheeze

Presentation

'This young patient has cystic fibrosis. They are cachectic and have central venous access. There is clubbing and evidence of diabetic pinpricks on the fingers, and coarse crepitations in both lower zones of the chest.'

- Comment on the patient's general appearance and look for any extra-pulmonary manifestations.

Investigations

- Diagnosis
  - Guthrie test
  - further genetic testing: ΔF508 is the commonest mutation
  - sweat test: sweat sodium and chloride >60 mmol/L
- Complications
  - chest: chest X-ray, high-resolution CT, sputum culture, spirometry
  - diabetes: oral glucose tolerance test
  - pancreas: faecal elastase
  - liver: ultrasound
  - sinuses: CT scan
  - bones: DEXA scan

Management

- MDT approach
  - doctor (CF specialist), CF specialist nurse, physiotherapist, dietician, clinical psychologist, GP, other medical teams (including diabetes)
- Specialist physiotherapy
- Antibiotics
  - see bronchiectasis section
- Bronchodilators
  - in those with airflow limitation
- Mucolytics
  - DNAse (nebulised)
- Nutrition
  - special diet
  - enteral feeding may be needed
- Management of extra-pulmonary complications, including diabetes

Questions

1 'How common is cystic fibrosis in the UK population? Please explain your answer.'
  - Cystic fibrosis occurs in approximately 1 in 2000 live births
  - Cystic fibrosis is an autosomal recessive condition
  - The chance of being a carrier in the UK is 1 in 22; the chance of two carriers mating is therefore approximately 1 in 500 ($22 \times 22 = 484$); the chance of both recessive alleles being passed on is 1 in 4; therefore, the incidence is 1 in 2000 (1 in 500×1 in 4)

2 'What organisms are commonly found in the sputum of patients with cystic fibrosis, and which are most important for prognosis?'
  - *Haemophilus influenzae*, *Staphylococcus aureus*, *Moraxella catarrhalis*, *Streptococcus pneumoniae*, *Pseudomonas aeruginosa*, *Burkholderia cepacia*, atypical mycobacteria, *Aspergillus fumigatus*
  - *Burkholderia cepacia* infection is a very poor prognostic sign

3 'List the extra-pulmonary manifestations of cystic fibrosis.'
  - Pancreatic: malabsorption, diabetes
  - Hepatobiliary: gallstones, cirrhosis
  - Intestinal: distal intestinal obstruction syndrome (meconium ileus equivalent)
  - Musculoskeletal: osteoporosis, arthritis
  - Sinusitis and nasal polyps
  - Male infertility

Key points

- Be able to recognise these patients, taking their age and general appearance into account.
- Recognise and comment on any extra-pulmonary manifestations.
- Be aware of common respiratory tract pathogens in cystic fibrosis.

Fibrotic lung disease

'Please examine this patient who is breathless.'

Findings

- General
  - dyspnoea, oxygen and walking aids
- Peripheral
  - cyanosis (peripheral and central), clubbing, signs of cor pulmonale
- Chest
  - scar (from biopsy), reduced expansion, fine end-inspiratory crackles

Presentation

'This patient has pulmonary fibrosis as evidenced by clubbing and fine inspiratory bi-basal crackles. The likely cause is systemic sclerosis as there is sclerodactyly, telangiectasia and microstomia.'

- This is a standard case, so look for a possible underlying cause:
  - features of rheumatoid arthritis: swan-neck and boutonnière deformities, Z-thumb, ulnar deviation at the wrist, nodules at the elbow
  - ankylosing spondylitis: 'question mark' posture
  - pacemaker and amiodarone facies
  - radiation burns and tattoos: radiation therapy
  - features of systemic sclerosis: sclerodactyly, telangiectasiae, beaked nose, furrowing of the mouth
- Also look for features of Cushing's syndrome (corticosteroid use)

Investigations

- Diagnosis
  - ABG (type 1 respiratory failure), pulmonary function tests, CXR, high-resolution CT thorax (ground-glass appearance, honeycombing)
- Cause
  - full history and examination, autoimmune screen, lung biopsy

Management

- Treat the underlying cause
- Pulmonary rehabilitation
- Corticosteroids
- Consider azathioprine and N-acetylcysteine in more advanced cases
- LTOT

Questions

1 'Please classify the causes of fibrosis into upper and lower zone.'
- Upper
  - berylliosis, radiation, extrinsic allergic alveolitis, ankylosing spondylitis, sarcoidosis, TB (mnemonic: 'breast')
- Lower
  - rheumatoid arthritis and other connective tissue diseases, idiopathic, drugs (methotrexate and amiodarone), asbestosis

2 'What would be the findings of full PFTs (including TLCO/KCO)?'
- Demonstrates a restrictive pattern (normal FEV_1:FVC, reduced FVC) with reduced TLC and TLCO

3 'What are the pulmonary manifestations of rheumatoid arthritis?'
- Lung fibrosis (which may also be secondary to methotrexate treatment)
- Pleural effusions
- Intrapulmonary nodules (including Caplan's syndrome)
- Obliterative bronchiolitis

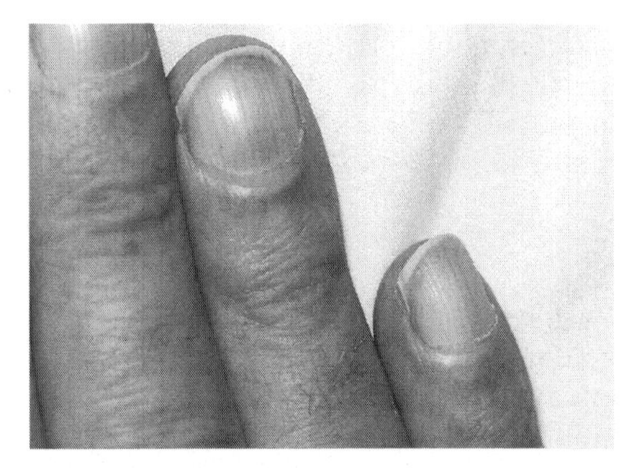

Figure 1.1 Finger clubbing

Key points

- Fine end-inspiratory crackles and clubbing are key findings.
- Know the causes of pulmonary fibrosis and the lung zones they affect.

Lung cancer

'Please examine this patient who has presented with a chronic cough.'

Findings

- General
  - cachexia, lymphadenopathy, hoarse voice
- Peripheral
  - clubbing, nicotine staining
- Chest
  - scars from previous lobectomy/pneumonectomy, radiation tattoos/burns, reduced chest expansion, tracheal deviation, reduced breath sounds, dullness to percussion, possible pleural effusion

Presentation

'This patient is cachectic and has finger clubbing and a hoarse voice. The trachea is deviated to the left and there is dullness and reduced breath sounds at the left lung base. The differential diagnosis includes a mass lesion and collapse.'

Look for the following:

- SVCO
  - raised JVP, oedematous face, distended veins over chest wall and neck, respiratory distress
- Metastases
  - hepatomegaly, bony tenderness, skin lesions, cervical lymph nodes
- Paraneoplastic disorders
- Pancoast's syndrome

Investigations

- Diagnosis
  - CXR, CT scan, PET scan
  - pleural fluid cytology
  - bronchoscopy and biopsy (including ultrasound-guided bronchoscopy)
  - CT-guided lung biopsy

- — biopsy of peripheral lesions
- — thoracoscopy

Management

- Referral to lung MDT team
- Dependent on staging and histology; surgical resection/chemotherapy/radiotherapy/palliation
- Surgical resection (pneumonectomy or lobectomy) is suitable for patients with adequate lung function and no medical contraindications
- Refer to patient's functional status using the WHO score.

Questions

1 'What are the four main histological types of lung cancer and how common are each of these?'
- — Lung cancer classification can be divided into small-cell lung cancer (SCLC) and non-small-cell lung cancer (NSCLC)
- — SCLC comprises of 20%–30% of all cases whereas NSCLC (squamous cell carcinoma (40%), adenocarcinoma (20%) and large cell carcinoma (10%–15%) account for the remainder of lung malignancies
- — Squamous cell carcinoma is the commonest form of lung tumour and has the closest correlation with cigarette smoking; it also has the best prognosis in terms of survival
- — Small-cell tumours have the worst prognosis as they have a rapid growth rate and metastasise early

2 'What paraneoplastic syndromes associated with lung cancer are you aware of?'
- — SCLC is the commonest form of tumour associated with paraneoplastic syndromes
- — Ectopic hormone secretion: ACTH (Cushing's syndrome), and ADH (causing SIADH – low sodium)
- — Lambert–Eaton's myasthenic syndrome (LEMS): a presynaptic syndrome (compared to myasthenia gravis which is a postsynaptic syndrome) caused by impaired release of acetylcholine from the nerve terminals due to production of antibodies to the calcium channels; it causes a proximal myopathy, hyporeflexia and an autonomic neuropathy

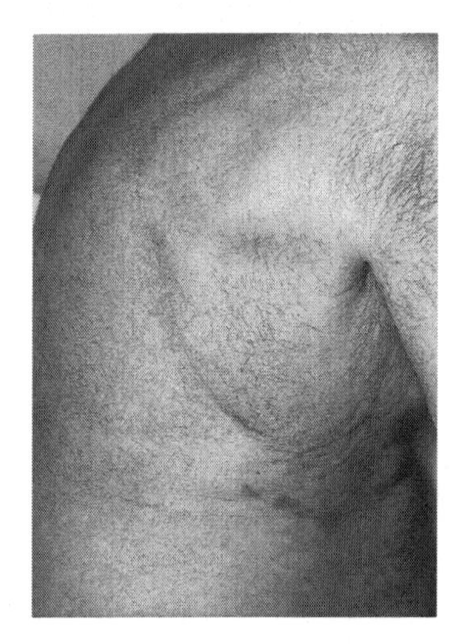

Figure 1.2 Thoracotomy scar with radiation tattoo

- Parathyroid hormone-related peptide release, resulting in hypercalcaemia, is associated with squamous cell carcinoma – note that hypercalcaemia is more frequently a result of bony metastases
- Hypertrophic pulmonary osteoarthropathy (HPOA): associated with squamous cell carcinoma; it results in gross finger clubbing and arthritis with radiological evidence of subperiosteal new bone formation

3 'What are the signs of Pancoast's syndrome?'

- Pancoast's syndrome is characterised by an apical lung tumour with involvement of the brachial plexus and cervical sympathetic nerves; patients may complain of pain in the shoulder/anterior chest wall, arm weakness, and have wasting of the intrinsic muscles of the hand and an ipsilateral Horner's syndrome (ptosis, miosis, anhidrosis and enophthalmos)

Key points

- Comment on any past treatment including surgery and radiotherapy.
- Comment on any associated complications as detailed above.
- Be able to differentiate between a pneumonectomy and lobectomy on clinical examination.

Old tuberculosis

'Please examine this gentleman's respiratory system.'

Findings

- General
  - may appear well or cachectic
- Chest
  - tracheal deviation, thoracotomy scar, rib resection, decreased breath sounds with reduced chest expansion, signs of fibrosis/ bronchiectasis, evidence of respiratory failure secondary to thoracoplasty

Presentation

'This gentleman appears well with no evidence of respiratory distress. He has a scar in the left supraclavicular fossa with fine inspiratory crepitations in the left upper zone. This is consistent with old TB.'

- Look for a supraclavicular scar from a phrenic nerve crush procedure, scarring from induced pneumothoraces, or a lateral thoracotomy scar.

Investigations

- CXR
  - raised hemidiaphragm on side of phrenic nerve crush procedure, upper lobe fibrosis (highly oxygenated area of lung, organism thrives), areas of cavitation
- CT scan
  - areas of fibrosis, loss of lung volume, thickened or calcified pleura
- Spirometry
  - post thoracoplasty may show an obstructive/restrictive defect

Management

- The patient may be completely well and not need treatment
- They may be investigated for possible recurrence of TB (bronchoscopy for a broncho-alveolar lavage) or manifestations of old TB such as pulmonary fibrosis

Questions

1 'Outline the current drug therapies available for treatment of TB, and their side effects.'
- The standard treatment for TB involves a combination of the following drugs: rifampicin, isoniazid, pyrazinamide and ethambutol; patients are initially treated with all four drugs for the initial two months of therapy followed by rifampicin and isoniazid for the subsequent four months; treatment is complete at six months
- Multidrug-resistant TB is that which is resistant to rifampicin and isoniazid
- Drug-induced hepatitis
  - rifampicin, isoniazid and pyrazinamide (check LFTs prior to commencement)
- Optic neuritis
  - ethambutol (visual acuity should be documented before starting treatment)
- Peripheral neuropathy
  - isoniazid (co-prescribe pyridoxine)

2 'Discuss past surgical treatments for TB.'
- Induced pneumothoraces
  - to collapse the affected lung; procedure repeated every few weeks
- Phrenic nerve crush
  - to paralyse the diaphragm and cause collapse of the underlying lung
- Plombage
  - insertion of polystyrene balls into the chest cavity to cause collapse of underlying lung
- Thoracoplasty
  - ribs around the infected cavity broken and pushed inwards to collapse underlying lung

3 What tests are used in the diagnosis of TB?
- CXR
- Sputum staining: Ziehl–Neelsen staining for acid-fast bacilli
- Bronchoscopy and washings
- Biopsies
- Whole-blood interferon or skin tests (Mantoux test)
- Always consider HIV testing in patients with suspected TB

Key points

- Know the past surgical treatments for TB and the associated clinical signs.
- Know the current treatment recommendations for TB.
- Know the serious side effects of anti-TB therapy.

Pleural effusion

'Please examine this patient's respiratory system. They have been complaining of a cough.'

Findings

- General
  - — dyspnoea, oxygen and walking aids
- Peripheral
  - — dependent upon cause
- Chest
  - — decreased expansion on inspection, decreased expansion on palpation, trachea deviated away from side of effusion, stony-dull percussion note on affected side, decreased breath sounds on affected side, bronchial breathing at limit of effusion

Presentation

'This patient has a right-sided pleural effusion as evidenced by reduced chest expansion, stony-dull percussion note and reduced breath sounds. The likely cause is lung cancer as the patient has clubbing, is cachectic and there is a radiation tattoo.'

- Search for a cause of the effusion:
  - — features of rheumatoid hands, butterfly rash (connective tissue diseases)
  - — cardiac disease: raised JVP and ankle swelling
  - — lung cancer: clubbing, wasting, Horner's syndrome, radiation scars, tattoos, lymphadenopathy
  - — liver disease: leuconychia, palmar erythema, spider naevi, gynaecomastia, parotid swelling
  - — renal disease: arteriovenous fistulae, scars from neck lines, sallow complexion, peritoneal dialysis catheters.
- Also look for signs of treatment:
  - — scars from pleural taps and chest drains.

Investigations

- CXR, pleural tap (should be US guided) sent for LDH, protein, pH (for empyema <7.2), amylase, glucose, cytology, microscopy and culture

Management

- Dependent upon underlying cause

Questions

1 'What differentiates an exudate from a transudate?'
  - An exudative effusion is defined by:
    - an effusion albumin:plasma albumin ratio >0.5
    - an effusion LDH:plasma LDH ratio >0.6
    - a pleural fluid LDH greater than 2/3 the upper limit of normal serum LDH (Light's criteria)

2 'What are the causes of an exudative and transudative effusion?'
  - Exudative
    - the 4 'I's: infiltration (neoplasm), infection, infarction (pulmonary embolus), inflammation (rheumatoid arthritis and SLE)
  - Transudative
    - cardiac failure, chronic renal disease, chronic liver disease (other rarer causes include Meigs's syndrome)

3 'List some drugs that may cause a pleural effusion.'
  - Amiodarone
  - Phenytoin
  - Methotrexate
  - Nitrofurantoin
  - Beta-blockers

4 'List two causes of low glucose in pleural fluid.'
  - Rheumatoid arthritis
  - TB

Key points

- Be able to differentiate between effusion and collapse (stony-dull percussion in an effusion).
- Know the causes of a pleural effusion and be familiar with Light's criteria.
- Look for an underlying cause for the effusion.

The patient with previous lung surgery

'This patient is complaining of a cough, please examine their respiratory system.'

Findings

- General
  - dyspnoea
- Peripheral
  - see extras section
- Chest
  - thoracotomy scar, reduced expansion on affected side, dull percussion note on affected side, decreased breath sounds on affected side

Presentation

'This patient has evidence of a right-sided pneumonectomy as evidenced by the lateral thoracotomy scar, reduced chest expansion and absence of breath sounds. The absence of breath sounds suggests that the procedure has been a pneumonectomy as opposed to a lobectomy.'

- Comment on the probable reason for surgery
  - lung cancer: clubbing, cachexia, Horner's syndrome, radiation scars and tattoos, lymphadenopathy
  - evidence of old TB: COPD (could have had operation for excision of large bullae and lung reduction surgery)
  - keyhole scars from video-assisted thoracoscopic surgery (VATS)
- Note: the other scenario in this station is that you have a patient who has had a lung transplant; if this is the case you may find that the only abnormality is a 'clam-shell' scar as illustrated below

Questions

1 'How would you differentiate between a lobectomy and a pneumonectomy?'
  - With a pneumonectomy (in adulthood) there are no breath sounds at all on affected side

— With a lobectomy there may still be some audible breath sounds

2 'What are the criteria for lung surgery in lung cancer?'
   — Patients must have an FEV_1 >1.51, a transfer factor >50%, no evidence of severe pulmonary hypertension and no evidence of metastatic disease (surgery is beneficial only in peripheral non-small-cell disease)

3 'What are the indications for a lung transplant?'
   — Patients with emphysema (usually with alpha-1-antitrypsin deficiency), idiopathic pulmonary fibrosis, idiopathic pulmonary hypertension, bronchiectasis, and cystic fibrosis may be considered for surgery

Key points

- Be able to recognise the range of thoracic scars and their underlying implications.

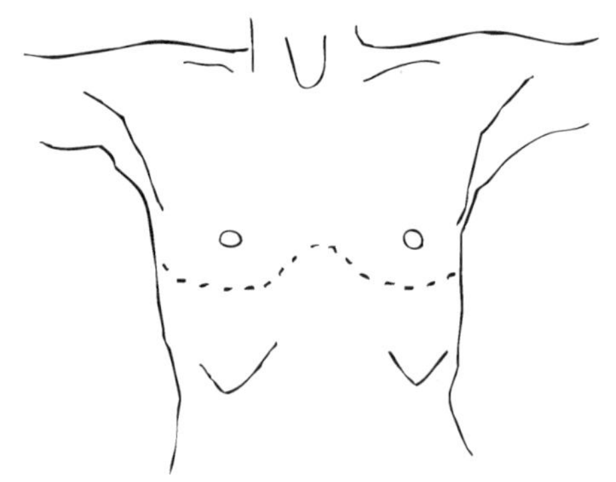

Figure 1.3 'Clam-shell' scar

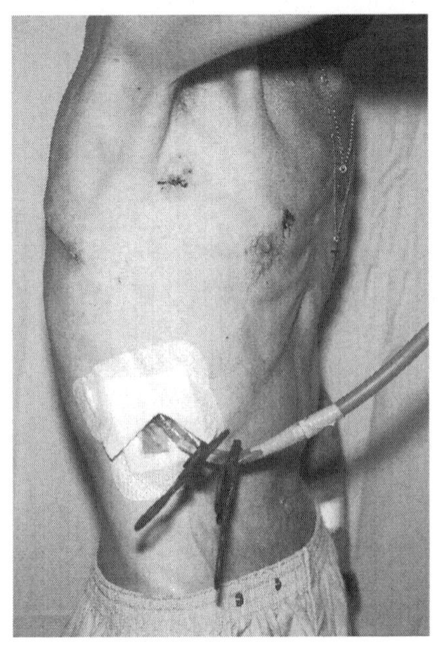

Figure 1.4 VATS and lateral thoracotomy scars

Respiratory station summary

- Observe the patient from the end of the bed and count the respiratory rate.
- Take a good look around the bedside and pick up on any clues that may help aid your diagnosis, e.g. sputum pot, different inhalers, oxygen mask.
- Ensure that you have time to examine both the front and the back of the chest in your examination – concentrate on the back as you are more likely to elicit findings.
- Be able to differentiate between consolidation and an effusion on examination.
- Look out for any scars which may aid you with the diagnosis.
- Know the contents of various inhalers.
- If you think the diagnosis is bronchiectasis and the patient has bi-basal coarse crackles, ask them to cough to see if the crackles clear.
- Be able to list the causes of upper and lower zone fibrosis.
- In cases such as pulmonary fibrosis and pleural effusion, look for an underlying cause for the pathology.
- If you see a surgical scar, look for clues such as a radiotherapy tattoo to help guide your diagnosis.
- Familiarise yourself with pulmonary function test results for common respiratory conditions such as COPD.
- Be aware of BTS, NICE and SIGN guidelines.

References and recommended reading

- Troosters T, Casaburi R, Gosselink R, *et al.* Pulmonary rehabilitation in chronic obstructive pulmonary disease. *Am J Respir Crit Care Med.* 2005; **172**: 19–38.
- BTS guidelines: www.brit-thoracic.org.uk/library-guidelines.aspx (accessed 9 December 2010).
  - Guidelines for non-CF bronchiectasis
  - Interstitial lung disease
  - Management of community-acquired pneumonia in adults
  - Guidelines for the radical management of patients with lung cancer
  - Guidelines for the management of pleural disease
- National Institute for Health and Clinical Excellence. *Chronic obstructive pulmonary disease: NICE Guideline CG101*. London: NIHCE; 2010: http://guidance.nice.org.uk/CG101 (accessed 9 December 2010).

Station 1: Abdominal

Contents

Hints for examination

- Read the candidate information carefully and look closely from the end of the bed for any signs that may help guide you to the specific examination; renal/haematology/liver/general abdominal.
- For a renal examination look for evidence of renal replacement therapy. Always check for an AV fistulae along the arm and look for fresh venepuncture marks at the fistula site. Also look for evidence of previous internal jugular lines or scars from peritoneal dialysis catheter insertions.
- For a liver exam, ensure that you comment on the presence of finger clubbing, Dupuytren's contractures and tattoos if present.
- Look in the oral cavity; gum hypertrophy in ciclosporin toxicity, poor dentition in chronic liver disease and glossitis in anaemia.
- If you find evidence of any cervical lymphadenopathy, always go on to check for axillary and inguinal lymphadenopathy.
- Ensure that the patient is adequately exposed and that they are lying as flat as possible prior to palpating the abdomen.
- Always make constant eye contact with the patient when palpating their abdomen to ensure that they are not in any discomfort. Apologise if you do cause any discomfort to the patient!
- Do not panic if you cannot elicit any abnormal findings; the patient may have a normal abdomen. In this case look for peripheral signs which may help guide you towards a diagnosis.
- If you elicit any hepatomegaly or splenomegaly, always try to comment on the size of it. In a case of splenomegaly, this may aid you with your differential diagnosis.
- Look for evidence of peripheral oedema.

An abdominal case with a normal abdomen

'Please examine this patient's abdomen.'

Findings

- General
  - cachexia
- Peripheral
  - oral/perioral telangiectasia, buccal pigmentation, hypertension, facial flushing, lymphadenopathy, cushingoid features
- Abdomen
  - no abnormality

Discussion

- If you finish your examination and have no positive findings it is likely that you are in one of the three following scenarios
  1. You have been given a patient with a normal abdomen
  2. You may have missed a peripheral sign that points to a diagnosis such as hereditary haemorrhagic telangiectasia, Peutz–Jeghers' syndrome, generalised lymphadenopathy, Cushing's syndrome or carcinoid syndrome
  3. You may have missed organomegaly

Presentation

Mention your findings and possible diagnoses:
'This patient has oral telangiectasia and looks clinically anaemic. There is nothing abnormal to find in the abdomen. The diagnosis may be hereditary haemorrhagic telangiectasia. I would like to know this patient's full blood count and ask if there is a family history of this disorder.'

Questions

1. 'What is the mode of inheritance of Peutz–Jeghers' syndrome?'
   - Autosomal dominant
   - This condition is characterised by mucocutaneous lesions predominantly on the face, oral mucosa and peripheries

— It also causes polyps within the gastrointestinal tract which may be complicated by bleeding, intestinal obstruction and malignant change

2 'How is a diagnosis of carcinoid syndrome established?'
   — Diagnosis is made by measurement of 24-hour urinary 5-hydroxyindoleacetic acid which is a degradation component of serotonin
   — Radiological methods may also be used to help identify the primary

3 'What is the mode of presentation of hereditary haemorrhagic telangiectasia?'
   — Recurrent epistaxis in childhood +/– family history
   — The patient is at risk of haemorrhage from AV malformations, particularly pulmonary and cerebral

Key points

- If you are unable to elicit any positive findings, you may have been given a 'normal abdomen' as your case.
- Ensure you look carefully for the presence of any peripheral signs to help guide your diagnosis.
- Do not make up any findings!

Chronic liver disease

'Please examine this patient who has presented with abdominal swelling.'

Findings

- General
  — cachexia, poor dentition, unkempt patient, muscle wasting, loss of body hair, gynaecomastia, testicular atrophy, tattoos, pedal oedema
- Skin
  — excoriation marks, purpura, spider naevi (most likely to be seen on the chest/back)
- Hands
  — clubbing, palmar erythema, leuconychia, Dupuytren's contractures
- Face
  — icterus, parotid swelling, pallor
- Abdomen
  — caput medusae, ascites, hepatomegaly, hepatic bruit, splenomegaly, evidence of previous ascitic taps/drains

Presentation

'On examination this patient has evidence of chronic liver disease. There is leuconychia and palmar erythema. The patient is icteric and there are excoriation marks. There is marked ascites with prominent superficial veins over the abdominal wall.'

- This is a common presentation, so be sure to look for an underlying cause; the most common cause is alcoholic liver disease (ALD)
- ALD
  — parotid swelling, unkempt, cachexia, Dupuytren's contracture
- Hepatitis
  — tattoos, injection sites
- Wilson's disease
  — Kayser–Fleischer's rings, neurological manifestations
- Haemochromatosis
  — slate-grey pigmentation, evidence of diabetes

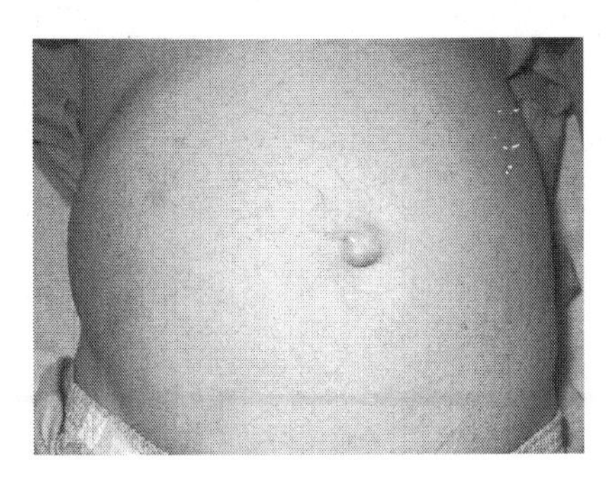

Figure 1.5 Evidence of ascites: prominent abdomen with distended abdominal wall veins

- Alpha-1-antitrypsin deficiency
  - shortness of breath, hyperinflated chest, clubbing

Investigations

- Bloods
  - FBC, U&Es, LFTs, coagulation screen, hepatitis serology, caeruloplasmin, ferritin, autoantibodies and immunoglobulins, AFP, TFTs, coeliac screen, alpha-1-antitrypsin levels
- Ascitic fluid
  - cell count (>250 white cells/mm^3 is indicative of spontaneous bacterial peritonitis), protein concentration, culture
- Imaging
  - ultrasound scan: check for hepato/splenomegaly, confirm ascites
  - Doppler flow studies of the hepatic/portal vein: rule out thrombosis
  - CT abdomen (adds little if USS is normal)

Management

- Dependent on the underlying cause
  - alcohol: alcohol avoidance, diuretics, vitamin B compound, thiamine, OGD to look for oesophageal varices
  - viral hepatitis: hepatitis C – interferon +/– ribavirin
  - Wilson's disease: penicillamine
  - haemochromatosis: venesection (to a ferritin level <50)
  - alpha-1-antitrypsin deficiency: supportive, advise not to smoke
- All these patients should be considered for liver transplantation on

an individual basis if severity dictates

Questions

1 'What are the signs of decompensated chronic liver disease?'
- Encephalopathy: grade 1 (altered mood/behaviour) to grade 4 (coma)
- Asterixis
- Jaundice
- Ascites
- Hepatic fetor
- Constructional apraxia: unable to draw a five-pointed star

2 'What are the reversible causes of hepatic encephalopathy?'
- Alcohol
- Drugs
- GI haemorrhage
- Infection
- Constipation

3 'What are the causes of ascites?'
- Transudative (protein level <30 g/dL):
  - chronic liver disease, nephrotic syndrome, congestive cardiac failure, hypoalbuminaemia, myxoedema, chylous, Meigs's syndrome
- Exudative (protein level >30 g/dL):
  - malignancy, pancreatitis, TB, Budd–Chiari's syndrome, portal vein thrombosis

4 'What are the possible indications for liver transplantation in an adult?'
- Acute causes
  - paracetamol poisoning (acetaminophen poisoning)
  - other drugs, e.g. isoniazid, phenytoin, sodium valproate
  - acute hepatitis
  - Epstein–Barr's virus
  - cytomegalovirus
- Chronic causes
  - alcoholic liver disease
  - primary biliary cirrhosis
  - primary sclerosing cholangitis
  - chronic viral hepatitis
  - Wilson's disease
  - Budd–Chiari's syndrome

- hepatic malignancy

5 'What criteria are used for liver transplantation in paracetamol poisoning?'
— The criteria used are 'The King's College criteria for liver transplantation', which include:
- lactate >3.5 mg/dL (0.39 mmol/L) four hours after early fluid resuscitation
- pH <7.30 or lactate >3 mg/dL (0.33 mmol/L) after full fluid resuscitation at 12 hours
- INR >6.5 (PTT >100s)
- creatinine >3.4 mg/dL (300 μmol/L)
- grade 3 or 4 hepatic encephalopathy

Key points

- Look out for examination findings which will guide you to the underlying diagnosis.
- In chronic liver disease many patients will have a shrunken cirrhotic liver rather than hepatomegaly.
- Be able to grade the various stages of hepatic encephalopathy.

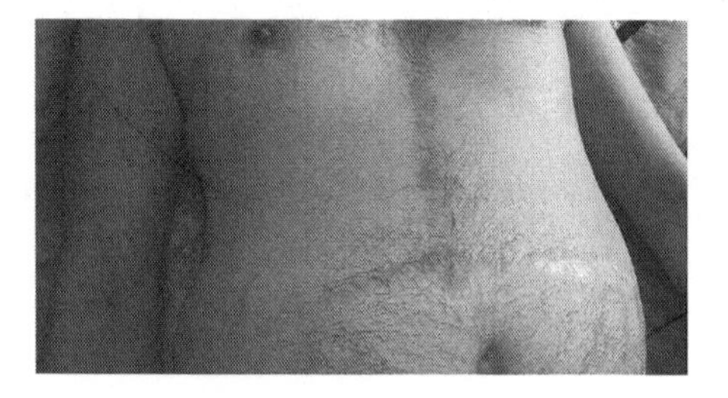

Figure 1.6 Scar seen post liver transplantation

Generalised lymphadenopathy

'This patient has presented with fatigue. Please examine their abdomen and any other relevant regions.'

Findings

- General
  - purpura, pallor, cachexia
- Peripheral
  - lymphadenopathy (cervical, supraclavicular, axillary), arthritis, enlarged tonsils
- Abdomen
  - splenomegaly, hepatomegaly, inguinal lymph nodes

Presentation

'This patient has marked lymphadenopathy. There are palpable nodes in the cervical, axillary and inguinal regions. There are purpuric lesions present on both arms. This could be a lymphoproliferative disorder with thrombocytopenia.'

- A difficult case, this could be picked up when examining the neck or following palpation of an enlarged spleen (and possibly liver).
- If following examination of the abdomen you feel that the patient could have a lymphoproliferative disorder, do not be afraid to re-examine the neck and axillae for nodes.

Investigations

- Diagnosis
  - blood tests (FBC, LDH, blood film, viral screen, autoimmune screen, LFTs), lymph node biopsy, bone marrow aspirate and trephine, chest X-ray and sputum for acid-fast bacilli
- Staging
  - CT chest/abdomen/pelvis, lumbar puncture

Management

- Dependent on the cause
- The objective of treatment of lymphoproliferative disease is to achieve a definitive cure or remission of disease where possible
- It is also important to manage symptoms accordingly

Questions

1 'What is the differential diagnosis of generalised lymphadenopathy?'
  - Lymphoproliferative disease: chronic lymphocytic leukaemia, acute lymphoblastic leukaemia, Hodgkin's and non-Hodgkin's lymphoma
  - Viral disease: includes HIV, EBV and CMV
  - Other infections: includes tuberculosis, brucellosis and toxoplasmosis
  - Inflammatory disease: sarcoidosis, rheumatoid arthritis, SLE

2 'How are Hodgkin's and non-Hodgkin's lymphoma differentiated pathologically?'
  - Through the presence of Reed–Sternberg's cells in Hodgkin's lymphoma (characteristic binucleate cells).

3 'What are 'B-symptoms' and what is their significance?'
  - These include weight loss (>10% in six months), unexplained fever >38.0°C and night sweats. B-symptoms are included in the Ann Arbor staging classification of Hodgkin's disease and indicate a worse prognosis.

4 'What are the complications of chronic lymphocytic leukaemia?'
  - Bone marrow failure: anaemia, thrombocytopenia (bleeding/bruising), leucopenia (infections)
  - Autoimmune haemolytic anaemia
  - Recurrent chest infections
  - Acute transformation (Richter's syndrome)

Key points

- When examining a seemingly normal abdomen do not forget to feel for inguinal lymph nodes.
- Be able to give a wide differential for lymphadenopathy.

Hepatosplenomegaly

'This patient has presented with purpura. Please examine their abdomen.'

Findings

- General
  - purpura, pallor, jaundice
- Peripheral
  - enlarged tonsils, lymphadenopathy (cervical, supraclavicular, axillary), features of chronic liver disease
- Abdomen
  - splenomegaly (note degree of enlargement), hepatomegaly, ascites, inguinal lymph nodes

Presentation

'This patient has 10 cm splenomegaly, with 4 cm hepatomegaly. There is no evident lymphadenopathy and no peripheral features of chronic liver disease. The most likely diagnosis is a myeloproliferative disorder.'

- A common case, the most likely causes are chronic liver disease and haematological malignancy; be sure to look for signs of both of these.
- After the enlarged liver and spleen have been palpated, re-examine for peripheral signs if necessary.

Investigations

- If the diagnosis is chronic liver disease, investigate as necessary.
- Otherwise, consider the following tests
  - blood tests: FBC, LDH, blood film, viral screen, autoimmune screen, LFTs
  - imaging: abdominal ultrasound
  - invasive: lymph node biopsy, bone marrow aspirate and trephine

Management

- Dependent on the cause

Questions

1 'What is the differential diagnosis of hepatosplenomegaly?'
   - Chronic liver disease
   - Lymphoproliferative disease: chronic lymphocytic leukaemia, Hodgkin's and non-Hodgkin's lymphoma (in acute lymphoblastic leukaemia the spleen is not usually greatly enlarged)
   - Myeloproliferative disease: chronic myeloid leukaemia (and acute myeloid leukaemia), polycythaemia rubra vera, essential thrombocythaemia and myelofibrosis
   - Viral disease: includes HIV, EBV, CMV and hepatitis B/C (may cause chronic liver disease)
   - Other infections: includes brucellosis, toxoplasmosis and leptospirosis
   - Inflammatory disease: sarcoidosis
   - Infiltrative disease: glycogen storage disease, amyloidosis

2 'What are the myeloproliferative disorders?'
   - A group of conditions caused by myeloid stem cell proliferation in the bone marrow.
   - The disorders are classified by the cell type that is proliferating
     - red blood cells: polycythaemia rubra vera
     - white blood cells: chronic myeloid leukaemia
     - platelets: essential thrombocythaemia
     - fibroblasts: myelofibrosis

3 'How would you diagnose glandular fever?'
   - Liver function tests may be deranged and a blood film may show atypical lymphocytes, though the most accurate tests are for heterophile antibodies (Paul–Bunnell's test, Monospot® test)

Key points

- Aim to decide between liver and haematological disease when giving a primary diagnosis.
- Look for clues of these two groups throughout the examination.
- Don't be afraid to re-examine for peripheral signs if necessary.

Multiple abdominal scars

'This patient has had an operation. Please examine their abdomen.'

Findings

- General
  - cachexia, cushingoid features
- Peripheral
  - oral ulceration, pallor
- Abdomen
  - multiple surgical scars (most commonly midline)
  - stoma sites (past/current)
  - mention that you would wish to perform a rectal examination
- Extras
  - gum hypertrophy and hypertension (side effects of ciclosporin)
  - extra-intestinal manifestations (*see* below)

Presentation

'This patient has inflammatory bowel disease as evidenced by cachexia, clinical anaemia and multiple scars on the abdomen suggesting fistulae, bowel obstruction or abscess drainage. A per rectal examination may be useful to look for fistulae and abscesses.'

- Mention scars and comment on nutritional state, side effects of therapy and extra-intestinal manifestations.

Investigations

- Stool culture
- Abdominal X-ray: rule out toxic megacolon
- Bloods including FBC and CRP
- Sigmoidoscopy/colonoscopy and biopsy (for histological confirmation)
- Barium studies looking for features of inflammatory bowel disease
- MRI enteroclysis (MRI of the small bowel)

Management

- Medical:
  - immunosuppression including steroids, 5-ASA and biological agents
- Surgical:
  - indications include fistulae, strictures and failure to respond to medical therapy
- Nutritional support and elemental and low residue diets
- Psychological support

Questions

1 'What are the differences between ulcerative colitis and Crohn's disease?'

Table 1.1 Differences between Crohn's disease and ulcerative colitis

	Crohn's disease	Ulcerative colitis
Distribution	Patchy ('skip lesions')	Continuous
Depth	Transmural	Superficial
Area involved	Whole gastrointestinal tract with predilection for terminal ileum and anus	Large bowel with predilection for rectum
Smoking	Higher risk	Lower risk
Fistulae and stenosis	Common	Rare

2 'What are the extra-intestinal manifestations of inflammatory bowel disease?'
  - Skin: erythema nodusum, pyoderma gangrenosum, aphthous ulceration
  - Joints: seronegative arthritides (mostly large-joint arthritis and sacroiliitis)
  - Eye: uveitis, episcleritis/scleritis, conjunctivitis
  - Hepatobiliary: primary sclerosing cholangitis (more likely in UC), cholangiocarcinoma (more likely in Crohn's)
  - Renal: oxalate stones

3 'What screening tests should you consider before initiating biological therapies?'
  - History and examination (looking specifically for features of tuberculosis)
  - Hepatitis serology

— CXR
— T-spot test (QuantiFERON® test is adequate if patient is not on steroids)

Key points

- Look for the extra-intestinal manifestations of inflammatory bowel disease to help guide the underlying diagnosis.
- Read up on key treatments available to treat these conditions.

Palpable kidneys

'This patient has presented with haematuria. Please examine their abdomen.'

Findings

- General
  - signs of anaemia, hypertension and renal replacement therapy
- Face
  - gum hypertrophy from immunosuppressants (see renal replacement therapy case), fundoscopy (hypertensive retinopathy)
- Abdomen
  - nephromegaly, bilateral flank masses, palpable cysts, hepatomegaly (associated with APKD), renal transplant and nephrectomy scars

Presentation

'The diagnosis is autosomal dominant polycystic kidney disease. On inspection there is the appearance of fullness in the flanks. On palpation there are bilateral bimanually ballotable masses in the flanks. It is possible to get above the masses and each moves with respiration.'

- Ensure you comment on any evidence of renal replacement therapy, e.g. the presence of a tunnelled vascular catheter, a peritoneal dialysis catheter or an AV fistula in the arms.
- In addition, on abdominal examination you may note a renal transplant scar in the iliac fossa and/or a subsequent midline scar from where the enlarged kidneys may have been removed.
- Indications for a nephrectomy in APKD include: recurrent infection, recurrent bleeding into cysts/rupture of cysts causing pain, significantly bulky kidneys.

Investigations

- Urinalysis
  - haematuria and proteinuria
- Ultrasound abdomen
  - gold-standard investigation

- Fundoscopy
  - hypertensive eye disease and evidence of Von Hippel–Lindau's syndrome
- CT abdomen
  - to look for malignancy if clinically indicated
- Echocardiogram
  - looking for mitral valve prolapse
- CT brain
  - looking for evidence of berry aneurysms associated with APKD
- Genetic testing
  - adult polycystic kidney disease (APKD 1 and 2)

Management

- Counselling
- Regular surveillance
- Monitor renal function
- Consider nephrectomy if appropriate
- Renal replacement therapy if required
- Investigation of first-degree relatives

Questions

1 'What is the genetic basis of APKD?'
  - Autosomal dominant APKD
    - 75% of patients have liver cysts by the 7th decade
  - Autosomal recessive APKD
    - liver cysts are a minor feature
  - Chromosomal mutations
    - Ch 1: PKD 1: accounts for 85% of cases; carries highest risk of developing ESRF
    - Ch 4: PKD 2: accounts for ~15% of cases; lowest risk of developing ESRF

2 'List some other manifestations of APKD.'
  - Cystic disease: liver, spleen, pancreas
  - Berry aneurysm: risk of SAH if ruptures (important to ensure good BP control)
  - Pain/haematuria: if ruptured cyst, stone, infection or renal cell carcinoma
  - Renal cell carcinoma: risk of malignant cyst transformation
  - Valvular disease: MVP, aortic valve disease
  - Gastrointestinal: colonic diverticulum formation, herniae

3 'List other renal cystic disorders.'
- Unilateral
  - benign renal cysts
  - renal cell carcinoma
  - polycystic kidney disease
- Bilateral
  - bilateral renal cell carcinoma
  - amyloidosis
  - Von Hippel–Lindau syndrome
  - tuberous sclerosis

Von Hippel–Lindau syndrome

- Autosomal dominant condition
- Defective tumour suppressor gene (acts by binding Elongin A)
- Features
  - angiomata develop in retina (may develop retinal haemorrhages), brain and spinal cord
  - angiomata develop in liver, kidney and pancreas
  - cerebellar haemangioblastoma (lateral lobes)
  - phaeochromocytoma
  - renal cell carcinoma

Key points

- Look for any scars that may indicate a nephrectomy.
- Look for any signs of renal replacement therapy.
- Know the causes of cystic renal disease.

Renal replacement therapy

'Please examine this patient's abdomen.'

Findings

- General
  - cushingoid appearance
- Face
  - corneal arcus, gum hypertrophy (secondary to ciclosporin)
- Neck
  - scars from internal jugular vein catheters, tunnelled-catheter exit site scar, parathyroidectomy scar
- Arms
  - AV fistulae (radiocephalic, brachiocephalic or brachiobasilic): feel for thrills, look for fresh venepuncture marks and calciphylaxis
- Abdomen
  - peritoneal dialysis catheter or scars from previous catheters
  - midline laparotomy scar or posterior subcostal scar: evidence of previous nephrectomy (possibly bilateral)
  - palpable mass in iliac fossa (graft may have been removed so mass may not be palpable, scars can be very faint)

Presentation

'This patient has had a renal transplant as evidenced by the presence of a renal transplant in the right iliac fossa. This is a functioning transplant as the patient does not have evidence of recent dialysis (fistula not recently needled/ no PD catheter present) or fluid overload.'

Comment on the following

- Renal replacement therapy
  - any other forms, e.g. vascular catheter, PD catheter, AV fistula (are they being actively used?)
- Fluid status
  - signs of fluid overload (raised JVP, peripheral oedema)
- Immunosuppression
  - ciclosporin: gum hypertrophy
  - corticosteroids: cushingoid features

Try to link with a possible underlying cause for renal failure

- Diabetes
  - injection sites on abdomen/lipodystrophy, pinprick marks on fingertips
- APKD
  - bilateral cystic masses in flanks
- Alport's disease
  - hearing aid

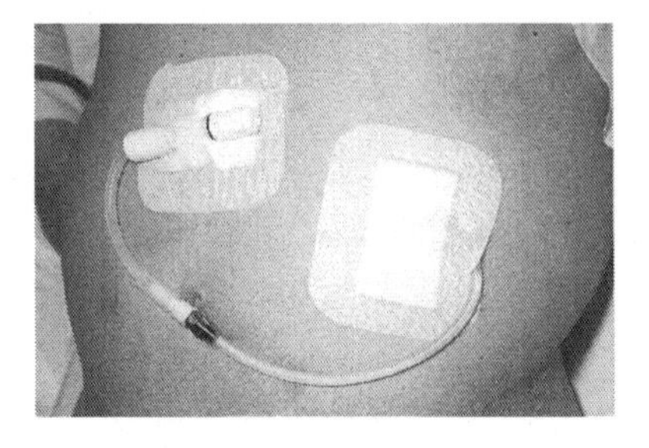

Figure 1.7 Peritoneal dialysis catheter

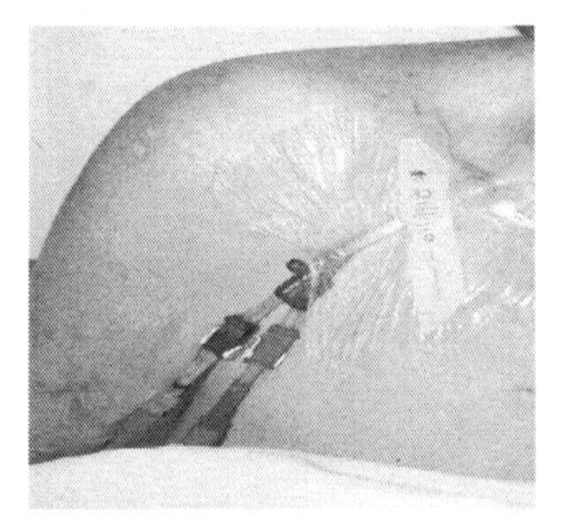

Figure 1.8 Vascular catheter

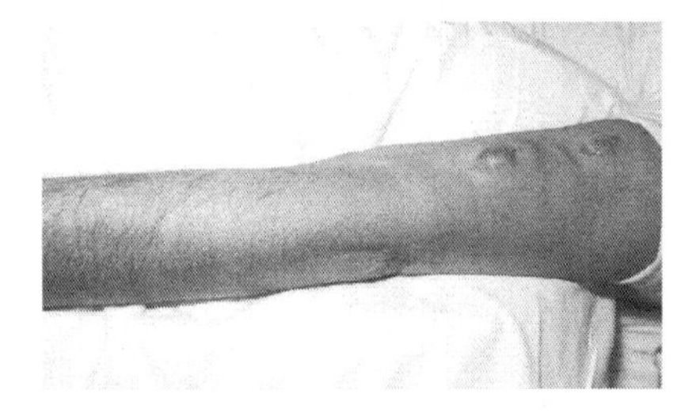

Figure 1.9 An arteriovenous fistula

Management

- Immunosuppression (including side effects)
  - tacrolimus (calcineurin inhibitor): diabetes and tremor
  - ciclosporin A: gingival hypertrophy, hypertension
  - mycophenolate mofetil: diarrhoea, nausea, vomiting
  - prednisolone: thin skin, easy bruising, muscle wasting, diabetes, hypertension, osteoporosis
- All except prednisolone give an increased risk of malignancy (yearly checks with dermatologist for signs of SCC and BCC)

Questions

1 'What are the commonest causes for a renal transplant?'
  - Diabetes mellitus
  - Glomerulonephritis
  - Adult polycystic kidney disease
  - Hypertension

2 'What are the signs of a failing transplant?'
  - Worsening renal function
  - Proteinuria
  - Tenderness over the transplant graft
  - Fluid overload
  - Interstitial fibrosis, tubular atrophy or vascular changes in a renal transplant biopsy

3 'Name some complications of haemodialysis.'
  - Cardiovascular disease, hypertension, anaemia, renal bone disease, increased infection rate, dialysis-related amyloidosis (B2-microglobulin accumulation), increased rate of malignancy, aluminium toxicity (now rare)

4 'Which diseases can recur after renal transplant?'
  - IgA nephropathy, focal segmental glomerulosclerosis (FSGS), membranous glomerulonephritis, mesangiocapillary glomerulonephritis, oxalosis, anti-GBM disease (rarely)

Key points

- Look closely for any scars which may represent a renal transplant.
- If there is a graft present, comment on how well it is functioning.
- Look for other forms of renal replacement therapy.
- Know the common causes for a renal transplant.

Splenomegaly

'This gentleman presented with symptoms of tiredness. Please examine his abdomen.'

Findings

- General
  - pallor, purpura
- Peripheral
  - lymphadenopathy
- Abdomen
  - enlarged spleen

Presentation

'This patient has evidence of splenomegaly as there is a palpable mass in the LUQ which moves inferomedially with respiration. I am unable to palpate above it, but can feel a splenic notch. On peripheral examination, there is evidence of pallor and bruising. There is no palpable lymphadenopathy.'

- Look for other signs indicating the underlying cause of the splenomegaly:
  - Felty's syndrome: rheumatoid hand signs
  - Portal hypertension: signs of chronic liver disease
  - Infective endocarditis: murmur, splinter haemorrhages
  - CML/CLL: lymphadenopathy (CLL), pallor
- Causes of splenomegaly
  - Congestive
    - cirrhosis, CCF, thrombosis of portal/hepatic or splenic veins
  - Haematological
    - lymphoma, leukaemia (acute and chronic), myeloproliferative disorders, myeloma, haemolytic anaemia, sickle-cell disease
  - Infection
    - schistosomiasis, malaria, leishmaniasis, EBV
  - Inflammatory
    - Felty's syndrome, SLE, rheumatoid arthritis
  - Infiltrative
    - Gaucher's disease, amyloidosis, Langerhan's cell histiocytosis, Niemann–Pick's disease

- Splenomegaly can be classified depending on the size (but remember that causes of massive splenomegaly may be moderate to start with!):
  - Moderate enlargement
    - rheumatological disease: rheumatoid arthritis, SLE, Sjögren's syndrome
    - infection: schistosomiasis, malaria, leishmaniasis, EBV
    - haematological: lymphoma, leukaemia, myeloproliferative disease, haemolytic anaemia
  - Massive enlargement
    - myelofibrosis
    - visceral leishmaniasis
    - chronic myeloid leukaemia

Investigations

- Bloods: FBC and blood film, LFTs, U&Es
- Biopsy: bone marrow aspirate, lymph node biopsy
- Imaging: USS abdomen, CT abdomen

Management

This is dependent on the underlying cause

Questions

1 'Differentiate splenomegaly from an enlarged kidney.'
  - Spleen
    - mass in the LUQ (moves towards RIF with respiration), unable get above a spleen on examination, dull to percussion, non-ballotable, presence of a palpable splenic notch
  - Kidney
    - moves minimally with respiration, can get above a kidney, resonant to percussion, ballotable, no notch palpable

2 'What advice should be given to a patient undergoing a splenectomy?'
  - Preoperative vaccinations: pneumococcal, meningococcal, *Haemophilus influenzae*
  - Lifelong prophylactic penicillin
  - Annual influenza vaccinations
  - Advice regarding seeking medical attention if unwell

3 'What are the symptoms of splenomegaly?'
  - Pain in the left upper quadrant
  - Pain referred to the left shoulder

- Early satiety (spleen pressing on stomach)
- If a massively enlarged spleen is grossly tender then consider splenic infarction as a possible cause

4 'What are the indications for a splenectomy?'
- Hypersplenism: autoimmune destruction of blood cells (for example, ITP)
- Mass effect of spleen
- Traumatic rupture
- Haematological malignancies (rarely)
- Congenital haemolytic anaemias: hereditary spherocytosis, elliptocytosis

Key points

- Be sure to measure splenomegaly to aid clinical diagnosis.
- Be sure to look for peripheral signs.
- Be able to categorise various causes of splenomegaly

Abdominal station summary

- In this station do not forget the normal abdomen. Remember the value of peripheral signs, e.g. lymphadenopathy; you may have your diagnosis before you lay your hand on the abdomen.
- Do not make up any findings!
- Don't forget to ask the patient if there is any pain in the abdomen before you begin palpation. This is basic but it can be embarrassing if you hurt the patient, and in all likelihood will result in a fail.
- Apologise if you cause the patient any discomfort.
- If possible, establish a cause for the patient's organomegaly to demonstrate to the examiners that you are thinking one step ahead.
- Demonstrate to the examiners that you are thinking ahead by inspecting for side effects of any treatments that the patient is receiving, e.g. the cushingoid patient with a renal transplant.
- Positioning is very important; ensure the patient is flat when you palpate the abdomen.
- Make your examination slick and quick as you may run out of time before you finish it, but at the same time don't be in a rush (a skill that can only be learnt with practice).
- Try not to be put off by unusual opening statements, for example, 'Please examine this patient with hypertension', (the opening statement will generally be relevant).
- If you palpate an organ, always ensure that you percuss it.
- Know the ways in which you can clinically differentiate a spleen from a kidney (this is a PACES classic).

Recommended reading

- Bailey B, Amre DK, Gaudreault P. Fulminant hepatic failure secondary to acetaminophen poisoning: a systematic review and meta-analysis of prognostic criteria determining the need for liver transplantation. *Crit Care Med.* 2003; **31**(1): 299–305
- Carter MJ, Lobo AJ, Travis SP; IBD Section, British Society of Gastroenterology. Guidelines for the management of inflammatory bowel disease in adults. *Gut.* 2004; **53**(Suppl. 5): v1–16.
- Wilkins BS. The spleen. *Br J Haematol.* 2002; **117**(2): 265–74.
- Update of guidelines for the prevention and treatment of infection in patients with an absent or dysfunctional spleen: www.bcshguidelines.com (accessed 10 December 2010).

Station 2:
History taking

Contents

Hints for the history-taking station

- The history-taking station is often thought of as being one of the easier stations; this misconception often costs candidates marks and sometimes their pass. You should not take this station lightly. You should see this station as an opportunity to excel, which may in turn give you breathing space if you have a bad station elsewhere.
- Make wise use of your five minutes before you enter the room; think about differential diagnoses and the pertinent question(s) you may ask to exclude/include them.
- Pay very close attention to what you are being asked to do. Commonly you will only be asked to find out more about the problem. You are probably not going to be expected to explain to the patient what you think is going on (unless the instructions specifically ask you to!), so focus on taking your history.
- You must always be polite and courteous to the patient.
- Do not forget simple things like introducing yourself.
- It is very important to establish a baseline; does the patient know why they have come to see you?
- Start with open questions as the patient will give you a great deal of information in the first three minutes; try not to interrupt them.
- Demonstrate to the examiners that you are organised by taking a structured history.
- Don't forget risk factors for your differential diagnoses.
- If the patient has brought a list of their medication, go through it with them; do not merely assume that it is up to date. Ask if there are any other medications? Ask about any specific allergies they may have.
- Be vigilant for any hidden agendas the patient may have (addressing their concerns specifically is a good way to flush this out!)
- Summarise appropriately and make it obvious that you are summarising so the examiners know.

- Remember to address the patient's ideas, concerns and expectations.
- Never forget the psychosocial element, as it is important when developing your problem list.

The patient with an abnormal blood result

Mr Warner is a 40-year-old male who on attendance at his well-man clinic was found to have an adjusted serum calcium of 2.75 mmol/L. All other blood tests were within normal parameters. He seems to have few symptoms. He is known to have hypertension and takes amlodipine 5 mg once daily; he is otherwise well. He is a factory worker, fully independent and lives with his wife and three children. He drinks 20 units of alcohol a week and does not smoke.

Examination revealed a clear chest and a soft and non-tender abdomen. A few small lymph nodes were palpated in his neck.

Key points for the patient

- You are 40 years old and as far as you are concerned, fit and healthy.
- You only went to the well-man check-up because your wife has been nagging you.
- You cannot think of any obvious symptoms, but on direct questioning you admit to having a troublesome dry cough.
- You have not coughed up any blood.
- You think you are becoming a little breathless on exertion.
- You have a good appetite, have not lost any weight and have not had night sweats.
- Your joints are not painful.
- Your GP felt something in your neck but you're not sure of the significance of this.
- You have no other symptoms beyond this.
- You have high blood pressure but no other medical problems. You take amlodipine, and have not recently used over-the-counter medications.
- You are not allergic to any medications.
- You do not smoke and never have done.
- You drink 20 units of alcohol a week.
- You cannot think of any specific family history; both your parents are alive and generally well.

Suggestions for the candidate

- Ensure you introduce yourself and check the patient's name and age.
- Ask the patient why they have come to see you.
- Ask the patient what they know about what has happened so far.
- Ask about any symptoms the patient may have.
- When the patient denies any symptoms be sure to clarify this. Proceed to direct questioning if necessary, working through a list of differentials.
- When finding out about the cough, cover other respiratory symptoms and consider both lung cancer and sarcoidosis as potential causes of the hypercalcaemia. Ask about constitutional symptoms of malignancy.
- Ask about arthralgia, as lack of this makes acute sarcoidosis less likely.
- Take a drug history, including over-the-counter medications, and ask about allergies.
- Ask about family history.
- Take an appropriate social history being sure to enquire about smoking.
- Be sure to summarise appropriately.
- Ask the patient about any particular concerns or expectations they might have.
- Explain to the patient that you cannot be sure of a diagnosis, but sarcoidosis is most likely (if asked directly about malignancy you cannot definitively rule it out at this stage).
- Explain to the patient that a chest X-ray and some special blood tests are required, and that you will see them next week with the results. You will likely need to do a biopsy of the lymph nodes.

Themes explored

- It is important to have a list of differentials for common abnormal blood test results (renal failure, hypercalcaemia and raised alkaline phosphatase among others).
- Tie into this list the symptoms (and signs) elicited. Lymphadenopathy points towards malignancy and sarcoidosis, and this link is strengthened with the presence of a dry cough.
- Look at the case as a whole; the young age, lack of social history and lack of constitutional symptoms in this patient swings the balance towards sarcoidosis.

Relevant information

Causes of hypercalcaemia

- Hyperparathyroidism (primary and tertiary)
- Malignancy (bony metastases, myeloma, paraneoplastic syndromes)
- Sarcoidosis
- Iatrogenic (vitamin D excess, milk-alkali syndrome)
- Dehydration
- Spurious result

The patient with back pain

Mr Quimby is a 78-year-old gentleman who complains of back pain. He has had this pain for four months, mainly in the thoracic region. There are no real exacerbating or relieving factors. He says it is worse at night. He is also complaining of fatigue and lethargy.

Examination is unrevealing. He is thin, has a soft, non-tender abdomen, normal heart sounds and a clear chest. A full blood count has found a haemoglobin of 10.9 g/dL with an MCV of 85 fL.

Key points for the patient

- You have had back pain in the middle of your back for four months and you feel it is getting worse.
- Your pain is worse at night-time.
- Your pain doesn't move anywhere and nothing makes it worse or better.
- You have been lethargic for some time now and are generally listless to the extent that your garden is overgrown as you don't have the energy to tend it.
- You don't weigh yourself but your clothes are slightly looser.
- You have had no change in bowel habit.
- You have no new urinary symptoms; you pass urine twice at night.
- You have not noticed any pains anywhere else.
- You have no chest symptoms. You do not sweat at night and have not had a fever.
- Your only past medical problem is that of prostate trouble and the GP gave you a tablet for that (you can't remember the name) around a year ago. Your urine stream is currently okay.
- To your knowledge you have not had any trauma to the back.
- You are a retired factory worker.
- You are an ex-smoker, stopping 25 years ago. You don't drink alcohol.
- You live with your son.
- With the lethargy you find it hard to have an active life. You used to play bowls but the back pain has put an end to this. You can still walk reasonably well.

- You are concerned that you may have cancer as your mother died of lung cancer.

Suggestions for the candidate

- Ensure you introduce yourself and check the patient's name and age.
- Ask the patient if they know why they have come to see you and what has happened so far.
- Start with an open question about his back pain.
- Find out where the pain is, its nature and if there is any radiation.
- Establish the onset, course and duration of symptoms.
- Establish if there are any exacerbating or relieving factors.
- Ask if there is any particular time when the symptoms are worse.
- Make sure you ask about red-flag symptoms: weight loss, systemic upset, night sweats, fevers and neurological symptoms.
- Screen for malignancy with your systems review.
- Take a thorough medical history.
- Enquire about family history.
- Ask about the patient's social situation and ensure you ask about the impact on their life.
- Be sure to summarise appropriately.
- Ask the patient about any particular concerns that they might have.
- Outline the investigations that are required, including blood tests and spinal imaging/bone scan.

Themes explored

- In a patient of this age group with back pain, malignancy is likely and needs to be asked about. Other red-flag symptoms are of relevance.
- The blood results here also point towards malignancy; be sure to take note of this. Knowledge is required of how to proceed with investigations.

Relevant information

Red-flag signs/symptoms for back pain

- Age <20 or >55
- Pain at night
- Constant/worsening pain

- Constitutional features of malignancy (weight loss, appetite loss, night sweats)
- History of malignancy
- Thoracic pain
- Pyrexia
- Bilateral lower limb neurological symptoms
- Sphincter disturbance

Causes of back pain in an elderly patient

- Degeneration (osteoarthritis)
- Vertebral collapse (osteoporosis)
- Solid malignancy metastases (a primary malignancy is less likely)
- Myeloma
- Paget's disease
- Trauma

Tumours most likely to metastasise to bone

- Breast
- Lung
- Prostate
- Kidney
- Thyroid
- Myeloma

The patient with a cough

Miss Usher is a 35-year-old female who has been troubled by a persistent dry cough for the last six months. She has tried a salbutamol inhaler and has stopped smoking though this has not led to an improvement in her symptoms. She has a peak flow of 440 L/min with no reversibility. Spirometry does not show an obstructive deficit.

There are no positive findings on examination. All her blood tests have been normal and a recent chest X-ray was reported as being 'clear'.

Instructions for the patient

- You are 35 and to your knowledge fit and healthy.
- You have had a cough for a long time; it is really starting to bother you.
- Your cough is worse at night-time.
- The cough is non-productive, with no blood produced.
- Your breathing is fine; you play badminton once a week and have noticed no change in your fitness levels.
- You are not wheezy.
- You have no chest pain or palpitations.
- You have not lost weight.
- You sleep with one pillow.
- On close questioning you experience some indigestion and a feeling of acid in your mouth. You do not have any abdominal pain.
- You have no medical history.
- You take the oral contraceptive pill and have not missed any doses. You are not on any other medications. You have no allergies.
- You do not think you are pregnant.
- Your mother has had a clot in her leg.
- You did not have any serious childhood illnesses.
- You are a lawyer. Your job is important to you and keeps you very busy. You work late commonly and often eat on the run, sometimes very late in the day.
- You used to smoke when you were out with friends but gave this up three months ago.
- You have no pets.

- You drink 35 units of alcohol per week.
- You are concerned that you have a serious illness.
- You are worried as the cough is keeping you awake at night, affecting your sleep and your performance at work.

Suggestions for the candidate

- Ensure you introduce yourself and check the patient's name and age.
- Ask the patient if they know why they have come to see you and what they know about what has happened so far.
- Start with an open question about her cough.
- Ask her specifically about when the cough is worse, and about sputum and blood production.
- Ensure you ask about other respiratory symptoms.
- Demonstrate to the examiners that you are working through your list of differentials (so specifically ask about symptoms of reflux, a common cause of cough).
- Check on medical history including childhood infections.
- As in any respiratory history take an occupational history and a smoking history.
- Enquire about pets and other common triggers for asthma.
- Take a drug history and be sure to ask about allergies.
- Ask about family history.
- Perform a systems' review (non-specific symptoms could be the only clue that the patient has lymphoma or sarcoidosis causing her cough).
- Ask about social situation and ensure you ask about impact on their life.
- Be sure to summarise appropriately.
- Ask the patient about any particular concerns that they might have (particularly relevant here).
- On making a likely diagnosis of reflux disease, offer reassurance and a trial of PPI, with follow-up in clinic in one month. Consider an OGD with regards to the duration of symptoms.
- Offer lifestyle advice in terms of reducing her alcohol intake and eating at sensible times.

Themes explored

- The final diagnosis may be from a different 'system' than the presenting complaint; be prepared to 'think outside the box'.

- It is important to know the 'red-flag' symptoms which may require urgent referral and endoscopic examination.
- Lifestyle advice may form a significant part of the management plan; this should be communicated to the patient.

Relevant information

Red-flag symptoms requiring urgent upper GI endoscopy

- Weight loss
- Dysphagia
- Symptoms despite treatment
- Symptoms >1 month
- Age >55

The patient with diarrhoea

Mr Wentworth is a 22-year-old male who complains of loose stools. He works as a car mechanic and is finding that his symptoms are disrupting his job.

Examination is unremarkable. He is thin, has a soft and non-tender abdomen, normal heart sounds and a clear chest. Preliminary blood tests have been sent, and he has had two stool samples that have come back negative on microscopy and culture.

Instructions for the patient

- You have had loose stools for six months.
- You generally open your bowels once a day.
- You have never noticed any blood or mucus.
- Your stools are difficult to flush and look greasy.
- You have lost one stone of weight in this period (this is not intentional).
- Your get a bloating sensation in your abdomen but do not have pain.
- You complain of fatigue.
- You suffer with regular mouth ulcers.
- You went to see your GP two years ago with a blistering and intensely itchy rash, which cleared of its own accord.
- You have a good appetite and do not suffer with night sweats.
- You have no medical history and do not take any medications.
- You are a smoker of 20 cigarettes/day.
- You don't drink alcohol.
- You find your symptoms embarrassing and they are interfering with your work.
- You have not had any foreign travel or had any infectious contacts.
- You eat a 'normal' diet and have not altered this at any time.
- You are not aware of any specific family history.

Suggestions for the candidate

- Ensure you introduce yourself and check the patient's name and age, as well as ask the patient why they have come to see you and what has happened so far.

- Start with an open question about his loose stools.
- Specifically ask about how often it occurs, about stool consistency and if there is any blood or mucus.
- Ask about abdominal pain, nausea and vomiting and weight loss.
- Enquire specifically about any extra-intestinal symptoms or associations of coeliac disease:
  - lethargy (anaemia)
  - dermatitis herpetiformis
  - aphthous ulcers
  - be aware of the predisposition to malignancy, particularly small-bowel lymphoma; ask about constitutional symptoms.
- In order to assist in ruling out irritable bowel syndrome, ask about anxiety, depression and ensure there is no evidence of other functional illness (for example functional dyspepsia).
- Enquire about family history of coeliac disease.
- Ask about foreign travel.
- Ask about infectious contacts.
- Always ask about alcohol intake where malabsorption is suspected.
- Ask about the patient's social situation and the impact of the symptoms on their life.
- Be sure to summarise appropriately.
- Ask the patient about any particular concerns that they might have.
- Outline the investigations that will be required (blood tests, possible upper GI endoscopy), and the management, given a diagnosis of coeliac disease (dietary change, referral to dietician).

Themes explored

- Coeliac disease is common in this age group and should be top of a list of differential diagnoses. Inflammatory bowel disease is less likely given the frequency of stool.
- It is important to ask about extra-intestinal manifestations and associations of coeliac disease; the history of dermatitis herpetiformis here helps to confirm the primary diagnosis.

Relevant information

Conditions associated with coeliac disease

- Dermatitis herpetiformis
- Hyposplenism
- IgA deficiency

- Type 1 diabetes mellitus
- Autoimmune thyroiditis
- Primary biliary cirrhosis

Common causes of malabsorption

- Coeliac disease
- Chronic pancreatitis (including cystic fibrosis)
- Carcinoma of the pancreas
- Crohn's disease (terminal ileal disease)
- Iatrogenic causes include small-bowel resection and radiation enteritis
- Infective causes are uncommon in the UK

The patient with jaundice

Mrs Valdes is a 64-year-old female with jaundice. She first noticed this after returning from India two months ago, since then it has gradually worsened. She also feels lethargic to the extent that she is struggling to make it through her days at work as a primary school teacher. She has a past history of hypothyroidism, for which she takes levothyroxine 150 mcg once daily and hypercholesterolaemia, for which she takes simvastatin 20 mg once daily. She is a moderate drinker, consuming one gin and tonic every other evening.

On examination she is icteric, she has palmar erythema and there are five spider naevi visible on her anterior chest wall. Her abdomen is soft and non-tender, and she has no palpable organomegaly. Her blood tests demonstrate deranged LFTs and an ultrasound scan of the liver has been arranged.

Instructions for the patient

- You first noticed the jaundice whilst you were in India but you have been feeling generally unwell for three months with tiredness and general lack of energy.
- Your urine is a normal colour and your motions have not changed.
- You have no abdominal pain, abdominal distension, nausea, vomiting, diarrhoea, fever or indigestion.
- You have not lost weight and your appetite is okay.
- You have severe itching that has been present for some time.
- You have hypothyroidism and take tablets for this. You are also on a statin for high cholesterol.
- You are fit and well otherwise.
- You work as a teacher and have done since university.
- Your husband is your only sexual partner.
- You were well throughout your stay in India, where you spent two weeks on holiday in Goa.
- You have a single gin and tonic every other evening.
- You have never been involved in IV drug use. You do not have any tattoos.
- You have never had any operations or blood transfusions.
- You are a non-smoker.

- Your mother has hypothyroidism and pernicious anaemia, and she takes steroids but you're not sure why.
- No one else in the family has been jaundiced.
- You live with your husband and two children.
- The GP has done some blood tests and organised a scan of your abdomen and sent you to see the specialist.
- You are worried it might be cancer.

Suggestions for the candidate

- Ensure you introduce yourself and check the patient's name and age.
- Ask the patient why they have come to see you and what has happened so far.
- Start with an open question about the jaundice.
- Ask specifically about the course of the illness and associated symptoms.
- Specifically enquire about the colour of the patient's urine and stool.
- Ask about other gastrointestinal symptoms.
- Demonstrate to the examiners that you are aware of the risk factors for jaundice by asking about alcohol, sexual history, intravenous drug abuse, blood transfusions, occupational history and travel history.
- Take a drug history and be sure to ask about allergies.
- Ask about family history and establish if there are any other symptoms of autoimmune disease in this patient.
- Ask about social situation and ensure you ask about impact on their life.
- Be sure to summarise appropriately.
- Ask the patient about any particular concerns and expectations that they might have.
- Communicate to the patient that further investigations are required. Medication can be offered to treat the itching.

Themes explored

- Painless jaundice is often caused by malignancy, though other causes need to be considered, especially in younger patients. Primary biliary cirrhosis is likely here; clues include the hypothyroidism and hypercholesterolaemia.
- It is important to establish the patient's concerns. Mrs Valdes is less likely to have a malignancy, but this cannot be ruled out until after imaging.

- Symptom control is important as well as making a diagnosis. Mrs Valdes' itching should be addressed.

Relevant information

Common autoimmune associations of PBC[1]

- Sjögren's syndrome
- Autoimmune thyroid disease
- Rheumatoid arthritis
- Systemic sclerosis
- Diabetes mellitus

Investigation of suspected PBC

- Deranged LFTs (raised ALP early in the disease, raised bilirubin later)
- Positive antimitochondrial antibody (M2 antibody is specific)
- Raised serum immunoglobulins, especially IgM
- Diffuse architectural change on liver ultrasound
- Hepatic granulomas, lymphocytic infiltrates and cirrhosis on liver biopsy

The patient with joint pains

Ms Leadbitter is a 34-year-old female who over the past two months has developed pains in the small joints of her hands. She is a secretary and is finding typing increasingly difficult. She also complains of stiffness in her fingers. She is a type 1 diabetic. Her blood sugars are well controlled. On examination she has tender joints in both hands but otherwise there is nothing abnormal to find.

Instructions for the patient

- You have pain in both of your hands, particularly the joints around the knuckles and the wrist.
- This pain started two months ago and you feel it is getting worse.
- It is an aching pain.
- You have noticed that your fingers and hands are also stiff in the morning. This stiffness goes away as the day progresses.
- None of your other joints are affected.
- You have been feeling more tired recently, but put this down to working extra hours.
- You have not lost any weight.
- You have had no problems with your eyes.
- You have no rashes and your nails are fine.
- You do not have chest pain, but you have noticed the beginnings of a dry cough. You have no problems with your breathing.
- On close questioning you do have a dry mouth.
- You are an insulin-dependent diabetic and are meticulous with control of your blood sugar.
- Other than the diabetes you are fit and healthy.
- You take insulin on a basal bolus regimen, simvastatin 40 mg once daily and ramipril 2.5 mg once daily.
- You do not have any allergies to medications.
- Your sister has pernicious anaemia.
- You live with your husband and three children.
- You work as a secretary for a law firm.

Suggestions for the candidate

- Introduce yourself and check the patient's name and age.
- Ask the patient why they have come to see you.
- Ask the patient what they know about what has happened so far (establish baseline).
- Establish which joints are affected; are the symptoms in both hands? Are any other joints are affected?
- Ask specifically about stiffness and clarify when it occurs.
- Ask about any lumps and bumps (nodules).
- Make sure you ask about symptoms of inflammatory bowel disease in order to help rule out an enteropathic arthritis.
- Ask about urinary symptoms to help rule out reactive arthritis (although more commonly an oligoarthritis or large-joint arthritis).
- Demonstrate that you know the associations of inflammatory arthropathies by asking about:
  - dry mouth and eyes (Sjögren's syndrome)
  - cough and dyspnoea (pleural effusions and lung fibrosis)
  - cardiac symptoms (pericarditis and pericardial effusions)
  - systemic symptoms (weight loss and lethargy)
  - rashes (psoriasis)
  - ophthalmic symptoms (episcleritis, scleritis and anterior uveitis).
- Ask about neurological symptoms in the hands, as carpal tunnel syndrome can develop in both rheumatoid arthritis and osteoarthritis.
- Take a thorough medical history.
- Take a thorough drug history and be sure to ask about allergies.
- Family history is important.
- Occupational history is important in this case.
- Try to ascertain the functional status of the patient.
- Be sure to summarise appropriately.
- Ask the patient about any particular concerns and expectations that they might have.
- Discuss a plan in terms of investigations and follow-up.

Themes explored

- This case demonstrates the link between autoimmune diseases, with the patient having an established diagnosis of type 1 diabetes. There is also a family history of pernicious anaemia.
- This case gives the candidate the opportunity to show that they know both the causes and associations of inflammatory arthritides.

- A key point is to distinguish early between inflammatory and mechanical arthritis.

Relevant information

Diagnostic criteria for rheumatoid arthritis (American Rheumatism Association 1987[2])

Four of the following seven criteria must be present, with the first four needing to be present for a period of over six weeks.

- Morning stiffness in and around joints lasting at least one hour before maximal improvement
- Soft tissue swelling (arthritis) of three or more joint areas
- Swelling (arthritis) of the proximal interphalangeal, metacarpophalangeal, or wrist joints
- Symmetric arthritis
- Rheumatoid nodules
- The presence of rheumatoid factor
- Radiographic erosions and/or periarticular osteopenia in hand and/or wrist joints

The patient with left arm weakness

Mrs Abedney is a 62-year-old female who three weeks ago developed left arm weakness which has subsequently not improved. She was fit and healthy prior to this. There is no family history of cardiac disease, she is a retired bank clerk and lives with her husband. She is a non-smoker.

On examination she has an upper motor neurone lesion affecting her left arm. Her blood pressure in clinic was 160/95. She has been told that she has had a stroke, and been commenced on aspirin, dipyridamole and ramipril. Her serum cholesterol and random glucose are pending.

Instructions for the patient

- You were fit and healthy up to this event.
- You are right handed.
- Three weeks ago you noticed some tingling in your left arm. Since then it has gradually become weaker and the whole of the left arm is now affected.
- The weakness is getting worse and not improving.
- The sensation in your left arm remains intact.
- Your right arm is normal and your legs are normal.
- Your speech has been normal.
- You have not experienced any headaches or visual symptoms.
- You have not sustained a head injury.
- Your swallowing is normal.
- Your bladder and bowel functioning has been normal.
- You have not lost any weight.
- You have been increasingly tired in the last three months.
- You have no urinary, respiratory, cardiac or gastrointestinal symptoms.
- You have no family history to speak of.
- You are a non-smoker and drink two units of alcohol every day.

- You saw your GP two weeks ago (one week after the onset of your symptoms) and she said you have had a stroke. She gave you some tablets and said she would refer you to a specialist.
- You are yet to have a scan of your head.
- The weakness has left you dependent upon your husband to manage the house, and he now has to help you get ready in the mornings.
- You are concerned that despite having treatment for two weeks, your symptoms have worsened.
- You feel a burden upon your husband and you hate asking for help. You feel as though you are losing your independence.

Suggestions for the candidate

- Ensure you introduce yourself and check the patient's name and age.
- Ask the patient why they have come to see you and what has happened so far.
- Ask the patient about their symptoms and establish a history of presenting complaint (you will note that the onset is gradual and the symptoms progressive, not in keeping with a stroke).
- Check the patient's handedness.
- Ask the patient about other neurological symptoms.
- Ensure you take a thorough medical history including asking about risk factors for stroke. Take a drug history.
- Ask about the patient's social situation including the impact of the symptoms on their life.
- Make sure that you do a thorough systems' review particularly asking about any symptoms that might suggest an underlying malignancy (especially if you have established there is a suggestion of weight loss and lethargy) or multiple sclerosis (ask about prior episodes of visual disturbance/incontinence/ataxia).
- Ask the patient about any particular concerns and expectations that they might have. Social support is important as well as making a medical plan for the patient.
- Explain to the patient that cerebral imaging is crucial here for diagnosis. Explain that you will expedite this and see the patient again later this week.

Themes explored

- The history leads you to think that the diagnosis is a space-occupying lesion.

- The diagnosis may be different from the GP's initial impression and from what the patient may have been told. Keep an open mind.
- Address the patient's concerns: explain that investigations are required to make a diagnosis, and that social support is available.

Relevant information

Causes of space-occupying lesions

- Benign tumour
- Malignant primary tumour
- Cerebral metastasis
- Colloid cyst
- Aneurysm
- Abscess
- Chronic subdural haematoma

Impact upon driving

In this case the patient should not drive until a diagnosis is made, at which point advice can be taken from the DVLA.

The DVLA's rules are quite complex. A craniotomy requires at least a six-month period off driving, with varying rules for benign tumours and at least 1–2 years off for malignant tumours after treatment.

The patient with visual disturbance

Mr Atherton is a 69-year-old male who has gradually developed blurred vision over the past three days. He also complains of a non-specific headache. He has a past history of hypertension, atrial fibrillation and hypercholesterolaemia, and had an OGD one year ago which demonstrated peptic ulcer disease.

On examination he looks generally well, his pulse is irregularly irregular, his blood pressure is 145/83, and his heart sounds are normal. Neurological assessment is normal. His GP thinks he is having transient ischaemic attacks. His medications are as below:

- Aspirin 75 mg OD
- Simvastatin 40 mg ON
- Bisoprolol 10 mg OD
- Candesartan 8 mg OD
- Lansoprazole 30 mg OD
- Quinine sulphate 300 mg ON

Instructions for the patient

- You have had this visual disturbance for a few days. Your vision is generally blurred and you cannot read the small print of the newspaper as well as you used to.
- The visual disturbance is getting worse.
- You have not had anything like this before.
- You don't wear glasses and last went to the optician one year ago.
- You have also had a headache over the same period.
- This headache hasn't particularly bothered you but it is not normal and you notice that it hurts when you wash your hair in the shower.
- You are also noticing pain when you chew.
- You don't have any floaters or flashing lights, or any symptoms of a 'curtain coming down'.
- You don't have any nausea or vomiting.
- There has been no recent head injury.

- Your legs, face and speech have been fine, though you have pain in both shoulders. Swallowing and bladder and bowel function are normal.
- Your doctor says you have high blood pressure, high cholesterol, an irregular heart beat and you take some tablets for all of these (you don't have any allergies).
- You had some abdominal pain last year and were found to have a stomach ulcer.
- You live with your wife who has dementia. You are her main carer.
- You worked as a banker in the past.

Suggestions for the candidate

- Ensure you introduce yourself and check the patient's name and age.
- Ask the patient why they have come to see you and what they know about what has happened so far.
- Ask an open question to start with about his blurring of vision.
- Ask about floaters, flashing lights, diplopia, eye pain and the use of glasses.
- Ask specifically about the headache and what precipitates it.
- Ask about other neurological symptoms.
- Show the examiners you are working through your differentials:
  - acute glaucoma (sudden onset, nausea and vomiting)
  - space-occupying lesion (early morning headache, worse on sneezing/coughing/leaning forward)
  - cerebrovascular disease (sudden-onset symptoms that improve with time, risk factors)
  - temporal arteritis.
- Once you establish the likelihood of temporal arteritis, demonstrate to the examiners that you know it is associated with polymyalgia rheumatica and ask about symptoms such as fatigue and stiffness.
- Take a thorough medical history as well as a drug history including allergies.
- Ask about family history.
- Ask about the patient's social situation and ensure you ask about the impact on their life.
- Be sure to summarise appropriately.
- Ask the patient about any particular concerns or expectations that they might have.

- Explain to the patient that whilst further investigations are required, treatment (corticosteroids) needs to commence immediately to prevent further deterioration.

Themes explored

- Have an open mind as to the diagnosis (this case is about ensuring you take your own thorough history and are not swayed by the provisional diagnosis from the GP).
- Associations are important; once you establish the likelihood of giant-cell arteritis, show the examiners that you know it is associated with polymyalgia rheumatica.
- A treatment plan should be communicated to the patient where prompt action is required.

Relevant information

Giant-cell arteritis

- A large-artery granulomatous arteritis
- It presents with the following symptoms
  - headache, either occipital or unilaterally over the temporal region
  - scalp/temporal tenderness
  - tenderness of the temporal/occipital arteries
  - jaw claudication
  - sudden visual loss
  - constitutional symptoms: malaise, fever
- ESR and CRP are generally high on testing; biopsy of the temporal artery may give a definitive diagnosis (skip lesions are common).
- Early treatment with high-dose corticosteroids is imperative. Corticosteroids can be tapered down but treatment for two years is usual

The patient with weight loss

Miss Shibib is a 29-year-old student who has lost three stone in weight over the last six months. The weight loss has been unintentional. She complains of associated reduced appetite and lethargy. She is a 1st-year nursing student and is finding it difficult to complete her studies. A systems' review was unremarkable. She has no significant medical history except for UTIs, for which she has received antibiotics and made good recoveries.

Examination reveals a thin female. Her abdomen is soft and non-tender, with a clear chest and normal heart sounds. Preliminary blood tests have been done but results are not yet available.

Instructions for the patient

- You have lost three stone in weight in the last six months, going from 12 to 9 stone.
- You have not been intending to lose any weight.
- You have noticed that you have become increasingly thirsty, to the extent that you take a jug of water to bed.
- You wake up twice at night to pass urine.
- You have noticed that you are passing more urine more often than before.
- You have had two courses of antibiotics in the last six months for urinary tract infections.
- You don't have any other symptoms.
- Prior to this you have been fit and healthy. You take no regular medications and have no allergies.
- Your mother has thyroid problems but there is no other family history.
- You are a nursing student.
- You don't smoke or drink.
- You are worried you may have diabetes.

Suggestions for the candidate

- Ensure you introduce yourself and check the patient's name and age.

- Ask the patient if they know why they have come to see you and what has happened so far.
- Start with an open question about her weight loss.
- Quantify the weight loss and associated symptoms.
- Demonstrate to the examiners that you are covering symptoms of potential causes of weight loss by performing a thorough systems' review and particularly looking for the following features:
  - inflammatory bowel disease: ask about diarrhoea (including blood and mucus), constipation, mouth ulcers, abdominal pain and perianal symptoms
  - malignancy (including haematological): ask about bruising, bleeding, night sweats, fever and do a thorough systems' review to unveil potential sites of malignancy
  - thyroid disease: irritability, eye symptoms, heat intolerance, tremor, sweating and palpitations
  - addison's disease: dizziness and abnormal pigmentation
  - diabetes mellitus: polyuria, polydipsia, nocturia and recurrent infections
  - inflammatory conditions: enquire about back pain, joint pain and rashes
  - chronic infections (including TB): ask about travel history and infectious contacts; ask about night sweats, haemoptysis and fever
  - psychiatric: enquire about the patient's mood.
- Ask about the patient's social situation and ensure you ask about impact of symptoms on her life.
- Be sure to summarise appropriately.
- Ask the patient about any particular concerns that she might have (particularly relevant here).
- Suggest that diabetes is certainly a possible cause, although investigations are required to diagnose this and rule out other causes.

Themes explored

- Autoimmune diseases are commonly linked, especially in the PACES setting.
- Weight loss is not always due to malignancy, especially in young patients.
- Cases with a wide differential can be covered in the time period, so don't become disheartened when given a non-specific symptom.

Relevant information

WHO criteria for diagnosis of diabetes mellitus

- Symptoms of hyperglycaemia **and** a raised venous glucose (fasting ≥7 mmol/L or random ≥11.1 mmol/L) **OR**
- Raised venous glucose on two separate occasions (fasting ≥7 mmol/L, or random ≥11.1 mmol/L, or oral glucose tolerance test two-hour value ≥11.1 mmol/L)

History-taking summary

- This station should be relatively simple and you should score highly, but you should NOT take it for granted; remain focused and target this station for scoring top marks (which will allow you some leeway on the harder stations).
- Do the simple things well, be polite and courteous and introduce yourself.
- Make good use of your five minutes before you enter the room; plan a structured approach to the case.
- Start with an open question as the patient may well give you the full history and all the salient points, if they don't then you will still have time to fill in the gaps.
- Try not to interrupt the patient, but if they are rambling you must do so subtly and politely; you must practise this beforehand.
- Remember to look for hidden agendas particularly if it is an apparently easy case (e.g. is it a chronic hypertensive who is non-compliant with medications because of impotence brought on by beta-blockers, but who is too embarrassed to tell you this).
- Use all of the time you are given. It would be embarrassing to have to sit in silence for five minutes if you finish early. Summarising is a useful tool as it will remind you of things that you still need to ask.
- Allergies are important to consider as you cannot manage your patient until you know if they are allergic to specific medicines.
- Read the instructions very carefully; do not start explaining the management plan unless it specifically asks you to. The likelihood is that they will only want you to take a history. The examiners may then ask you what your approach will be when they question you.
- An easy station to practise – you should do it in groups with one person being the patient, one being the candidate and the others being the examiners.

References and recommended reading

1 Watt FE, James OF, Jones DE. Patterns of autoimmunity in primary biliary cirrhosis patients and their families: a population-based cohort study. *QJM*. 2004; **97**: 397–406.

2 Arnett FC, Edworthy SM, Bloch DA, *et al.* The American Rheumatism Association 1987 revised criteria for the classification of rheumatoid arthritis. *Arthritis Rheum*. 1988; **31**(3): 315–24.

■ National Institute for Health and Clinical Excellence. *Dyspepsia: management of dyspepsia in adults in primary care: NICE clinical guideline 17*. London: NIHCE; 2004. http://guidance.nice.org.uk/CG17/QuickRefGuide/pdf/English (accessed 10 December 2010).

Station 3: Neurology

Contents

Hints for the neurology station

- Ensure you pay close attention to the lead-in statement as this will guide your examination. Once you have decided upon a system be time efficient with this.
- Neurology cases lend themselves to being spot diagnoses (e.g. myotonic dystrophy, Parkinson's disease, old polio, and myasthenia gravis) therefore be alert from the moment you walk into the room.
- Looking around the bed is very relevant in neurology (e.g. callipers, walking aids, devices for measuring FVC).
- If you spot an obvious deformity (claw hand, wrist drop, Charcot's joint), be sure to direct your examination towards this.
- When commenting on power, use the MRC scale to grade it.
- Neurology is often thought of as being a particularly difficult station, but if you work on the premise that you are trying to find the site of the lesion (e.g. cerebrum, cerebellum, brainstem, spinal cord, anterior horn, peripheral nerve or muscle) and then consider the differential diagnosis, you will demonstrate to the examiners that you are organised and pragmatic.
- Have set routines for examining the cranial nerves and the peripheral nervous system (upper limb and lower limb) but be prepared to adapt your examination (e.g. you may be asked to examine just cranial nerves III-VII). Neurology particularly lends itself to this type of adjustment.
- You will earn yourself extra marks if you look for the underlying pathology, e.g. blood glucose pinprick marks, granuloma annulare, and cataracts in a case of peripheral neuropathy indicate that diabetes is the underlying cause.
- If you are unable to elicit reflexes you must demonstrate that you have attempted to elicit them with reinforcement.
- When assessing sensation, be sure to check for normal sensation on the chest first. Compare normal sensation to the peripheral sensation (i.e. ask if the sensation is different to the sensation on the chest, not merely 'Can you feel it?').
- It is important to decide early whether to assess sensation in a

dermatomal or a peripheral ('glove and stocking') pattern.

- Although the patients may have been examined many times and will be well versed in the routines, you must give clear instructions about what you want them to do.
- At the end of the examination ask to examine other relevant neurological areas (including fundoscopy).
- If you believe that the diagnosis is related to vascular disease then also mention that you would like to examine the cardiovascular system for risk factors.

Cerebellar syndrome

'Please examine this patient who has had problems with his balance, and recurrent falls over the past six months.'

Findings

- General
  - — intention tremor
- Neurological
  - — ataxic gait, truncal ataxia, hypotonia, pendular reflexes, dysmetria, dysdiadochokinesis, nystagmus, slurred speech, heel-shin ataxia
- Extras
  - — cause: features of multiple sclerosis (spastic paraparesis, sensory disturbance, INO)
  - — features of chronic alcoholism (chronic liver disease)
  - — Friedreich's ataxia (young patient, wheelchair/walking aids, pes cavus, kyphoscoliosis, absent ankle jerks with up-going plantars)

Presentation

'This man has cerebellar dysfunction. He has an ataxic gait with nystagmus and slurred speech. He also has multiple purpura. The most likely cause is excess alcohol consumption.'

- In this common case it is essential to look for a cause as well as to demonstrate the clinical signs. It is important to discuss functional status.
- Causes of cerebellar syndrome:
  - — multiple sclerosis
  - — alcoholic cerebellar degeneration
  - — posterior fossa space-occupying lesion
  - — brainstem vascular lesion
  - — inherited ataxias (e.g. Friedreich's)
  - — paraneoplastic syndromes
  - — drugs (phenytoin).

Investigations

- Dependent upon the presumed cause of cerebellar signs
- If multiple sclerosis is suspected
  - cranial and spinal imaging (MRI)
  - LP (oligoclonal bands and protein in CSF)
  - visual-evoked potentials

Management

- Dependent upon the cause of cerebellar syndrome

Questions

1 'How would you identify the site of a cerebellar lesion from clinical findings?'
  - The cerebellum is divided into a midline vermis and two cerebellar hemispheres
  - Disease of the vermis leads to truncal ataxia and ataxic gait
  - Disease of a hemisphere causes ipsilateral dysmetria, dysdiadochokinesis, an intention tremor, and fast-beat nystagmus towards the lesion
  - Multiple sclerosis (demyelination) causes a global deficit

2 'What is Friedreich's ataxia?'
  - Autosomal recessive disorder
  - A trinucleotide repeat on chromosome 9
  - Degeneration of the spinocerebellar tract causes cerebellar signs
  - Corticospinal tract damage and peripheral nerve degeneration lead to absent ankle jerks with extensor plantars
  - Pes cavus, scoliosis and diabetes are common features; other features include cardiomyopathy, cataracts and sensorineural deafness

Key points

- Prepare a specific examination for the cerebellar syndrome.
- When the cerebellar syndrome has been adequately demonstrated, be sure to look for possible causes.

Hemiparesis

'This patient has developed weakness. Please examine their upper limbs.'

Findings

- General
  - walking aids, PEG tube, wasting/oedema on affected side, upper limbs held in flexion, lower limbs held in extension
- Peripheral
  - bruising (on warfarin), amiodarone facies, irregularly irregular pulse
- Neurological
  - increased tone
  - reduced power (use MRC grade for power)
  - hyperreflexia and extensor plantars
  - decreased sensation
  - hemianopia
- Extras
  - observe for UMN facial nerve lesion
  - note any dysphasia/dysarthria
  - neglect/visual or sensory inattention
  - mention that you would like to perform a full cardiovascular examination

Presentation

'This patient has a right-sided upper motor neurone lesion affecting arm, leg and face. The likely cause is a stroke but I would include space-occupying lesion and demyelination in my differential diagnosis. He uses a frame to walk. I would like to examine his cardiovascular system including blood pressure readings, in order to look for a predisposing cause.'

- Comment on the side of stroke and the areas involved.
- Comment on any obvious underlying predisposing factors.
- Comment on functional status.

Investigations

- Neuroimaging (CT/MRI)
- Blood pressure
- Fasting glucose/cholesterol
- ECG/24-hour tape
- Echocardiogram
- Carotid artery Doppler

Management

- Thrombolysis (within 4.5 hours of onset of symptoms)
- Aspirin +/– dipyridamole
- Use a tool such as the NIH Stroke Scale to assess patient functionality at onset of stroke and at regular intervals after onset
- Speech and language assessment
- Management of predisposing factors and secondary prevention
- Rehabilitation (OT and physiotherapy)
- Psychological and nutritional support
- Measure progress with tools, e.g. Modified Rankin's score to assess prognosis, and length of hospital stay

Questions

1 'What extra investigations are merited in young patients with proven stroke?'
   - Thrombophilia screen
   - Homocystine levels (postulated to promote atherosclerosis leading to stroke)
   - Bubble echo (looking for a patent foramen ovale)

2 'What score may help guide a decision to warfarinise a patient with atrial fibrillation?'
   - CHADS2
     - C: congestive cardiac disease
     - H: hypertension
     - A: age greater than 75
     - D: diabetes mellitus
     - S: stroke (scores 2)
   - A score of 2 points or more suggests a high risk of stroke and is an indication for anticoagulation with warfarin
   - A score of 1 suggests a moderate risk of stroke and anticoagulation with aspirin or warfarin is advised (depending on patient preference)

- A score of 0 is low risk and anticoagulation with aspirin alone is advised

3 'What is 'lateral medullary syndrome'?'
- This is also known as Wallenberg's syndrome; it is caused by a brainstem stroke in the territory of the vertebral or posterior inferior cerebellar artery
- Clinical features include:
  - ipsilateral signs: Horner's syndrome, nystagmus, facial sensory impairment, ataxia and diplopia
  - contralateral signs: pain and temperature loss over opposite arm and trunk (spinothalamic tract)

4 'What is the Bamford stroke classification?'
- This system localises strokes anatomically on the basis of clinical findings, and provides information on prognosis. Stroke is split into four categories:[1]
- Total anterior circulation infarction (TACI)
  - higher cortical dysfunction
  - homonymous hemianopia
  - unilateral motor and/or sensory deficit affecting at least two of the three areas: face, arms and legs
- Partial anterior circulation infarction (PACI)
  - two of the three features of TACI, or isolated higher cortical dysfunction, or isolated unilateral neurology
- Lacunar anterior circulation infarction (LACI)
  - pure motor or pure sensory deficit, or ataxic hemiparesis
- Posterior circulation infarction (POCI)
  - cranial nerve deficit with contralateral hemiparesis or sensory deficit, or bilateral stroke, or disorders of conjugate eye movement, or isolated homonymous hemianopia, or isolated cerebellar deficit

Key points

- Establish the diagnosis and the cerebral area(s) affected.
- Mention to examiners that you would like to examine the cardio-vascular system (for risk factors).
- Look for evidence of disability caused by the hemiparesis and comment upon this.

Myotonic dystrophy

'This patient who has recently had cataract surgery has presented with ptosis. Please examine their face, and proceed.'

Findings

- General
  - 'Myopathic facies', wasting of sternocleidomastoid muscle group, speech (nasal or dysarthric), pinprick marks on fingertips from blood glucose testing
- Face
  - Bilateral ptosis, wasting of facial muscles with hollowing of temporal fossae and cheeks, frontal baldness, smooth forehead, cataracts
- Hands
  - Generalised weakness and wasting of upper limbs
- Key points: the candidate must be able to demonstrate myotonia
  - Grip myotonia
    - ask the patient to clench their fist as tightly as possible and release; this will demonstrate a slow release instead of rapid finger extension
    - this can also be demonstrated by shaking the patient's hand, which will demonstrate slow release of grip
  - percussion myotonia
    - this can be demonstrated by using a tendon hammer to tap the thenar eminence, which will display a muscle twitch followed by a slow relaxation of the muscle group

Presentation

'On examination, this patient has evidence of myotonic dystrophy. This is demonstrated by the myopathic facies: elongated face, wasting of temporal muscles, frontal balding and bilateral ptosis. I have demonstrated evidence of myotonia as there was slow release of grip after shaking his hand and percussion myotonia on tapping the thenar eminence.'

- Comment on any evidence of complications (see below).

Diagnosis

- Electromyography
- Muscle biopsy
- Genetic analysis

Management

- Reduction of myotonia using drugs such as phenytoin, mexiletine and carbamazepine (these drugs are sodium channel blockers; they reduce myotonia but may increase weakness)
- Identification and treatment of associated complications (see below)
- Genetic counselling
- Avoidance of general anaesthesia where possible

Questions

1 'List some complications associated with myotonic dystrophy.'
  - Cardiac
    - dilated cardiomyopathy, cardiac arrhythmias
  - Respiratory
    - risk of aspiration due to muscle weakness; may require non-invasive ventilation due to myotonia affecting respiratory muscles
  - Gastrointestinal
    - dysphagia, delayed gastric emptying
  - Endocrine
    - increased risk of diabetes, thyroid dysfunction
  - Reproductive
    - testicular atrophy, infertility
  - Other
    - cataracts

2 'What is the genetic basis of this condition?'
  - Autosomal dominant
  - Trinucleotide-repeat disorder showing genetic anticipation (expansion of an unstable CTG trinucleotide repeat in the myotonic protein kinase gene)
  - Gene located on chromosome 19

3 'What problems are associated with general anaesthesia?'
  - Sedatives and neuromuscular blocking drugs may lead to cardiorespiratory complications and delayed recovery from anaesthesia

- Depolarising neuromuscular blocking agents should be avoided, e.g. suxamethonium

4 'What changes would be detected on electromyography (EMG) in myotonic dystrophy?'
- Electromyography shows the electrical potential generated by muscles when they are neurologically or electrically activated
- Myotonic dystrophy produces high-frequency activity that varies, producing a whining sound on the loudspeaker ('dive-bomber')

Key points

- Try to identify myotonic dystrophy early from the characteristic facial changes.
- Be sure to examine for specific signs of myotonia (grip and percussion).
- Look for evidence of extra-neurological complications.

Ocular palsies

'Please examine this patient's eyes as they have been complaining of double vision.'

Note: The term 'ocular palsy' refers specifically to the loss of function of an ocular muscle due to pathology in the nerve supplying it. We have also considered other causes of diplopia in this section.

Findings

- General
  - walking aids, eye patch, ptosis
- Neurological
  - third (oculomotor) nerve palsy: complete ptosis, eye looks down and out, pupil may be dilated and unreactive dependent on the cause
  - fourth (trochlear) nerve palsy: causes weakness of downward movement of eye, causing vertical diplopia (rare)
  - sixth (abducens) nerve palsy: inability to abduct affected eye
  - complex ophthalmoplegia: a combination/no specific nerve involvement
  - internuclear ophthalmoplegia: impaired adduction, unilaterally or bilaterally
- Extras
  - stigmata of diabetes mellitus
  - myasthenia gravis: bilateral ptosis, fatiguability
  - Graves's disease: proptosis, neck lump/scar
  - multiple sclerosis: spastic paraparesis, cerebellar signs
  - mitochondrial diseases: hearing aid, proximal myopathy, ataxia, pacemaker (cardiomyopathy)
  - Miller Fisher's syndrome: peripheral neuropathy, ataxia, areflexia

Presentation

'This patient has a complex external ophthalmoplegia as evidenced by diplopia in directions of gaze that are not attributable to a single nerve lesion. The likely cause is myasthenia gravis as this patient also has bilateral ptosis, demonstrable fatiguability and a midline thoracotomy scar which would be consistent with a prior thymectomy.'

- Where possible try to work out which nerve is the culprit, though bear in mind that the patient may have a complex ophthalmoplegia.

Investigations

- Neuroimaging if a nerve lesion is suspected; MRI is most helpful as it is important to obtain good views of the brainstem and posterior fossa
- Investigations for causes of mononeuritis multiplex
- Investigations for myasthenia gravis and thyroid disease in a complex ophthalmoplegia

Management

- Dependent upon the underlying cause

Questions

1 'What are the causes of an oculomotor nerve palsy?'
   - 'Surgical' – these causes generally affect the pupil
     - posterior communicating artery aneurysm
     - space-occupying lesion in midbrain/sphenoid wing/near cavernous sinus
     - haemorrhage
   - 'Medical' – these causes often do not affect the pupil
     - causes of mononeuritis multiplex
     - demyelination
     - infarction

2 'What are the causes of an abducens nerve palsy?'
   - Causes of mononeuritis multiplex
   - Vascular lesion
   - Malignancy
   - Demyelination
   - Infection (Lyme disease, syphilis)
   - Raised intracranial pressure ('false localising' sign)
   - Wernicke's encephalopathy can cause a bilateral abducens nerve palsy

3 'What are the causes of a complex ophthalmoplegia?'
   - Nerve lesions: demyelination, mononeuritis multiplex
   - Neuromuscular junction: myasthenia gravis
   - Muscle: Graves's disease
   - Mitochondrial disease

4 'What are the causes of an internuclear ophthalmoplegia?'
 — Multiple sclerosis
 — Vascular disease

Key points

- Attempt to work out if the ocular palsy is due to a specific nerve lesion. If this is not possible, consider a complex ophthalmoplegia.
- The pupil may give some idea towards aetiology in an oculomotor palsy.

Parkinson's disease

'Please examine this patient's gait and proceed. They have presented with unsteadiness.'

Findings

- General
  - resting tremor (4–6 Hz), mask-like face, dyskinesias
- Neurological
  - hypertonia (rigidity), bradykinesia, festinant gait, soft and monotonous speech, lack of sweating, micrographia; note that the features of parkinsonism often worsen with concurrent activity (for example, waving the right arm when assessing for hypertonia in the left)
- Extras
  - treatment: apomorphine infusion (subcutaneous)
  - full assessment should also include mood, cognition and functional status

Presentation

'This patient has evidence of Parkinson's disease. There is a resting tremor that worsens with concurrent activity, as well as rigidity and bradykinesia. I note an apomorphine infusion being administered via a syringe driver.'

- This is a common presentation at the neurology station and should be recognisable from the classic triad of parkinsonism: bradykinesia, rigidity and a resting tremor. Be sure to discuss the functional status of the patient and consider causes other than just Parkinson's disease.
- Causes of parkinsonism:
  - Parkinson's disease
  - drugs (neuroleptics)
  - CVA or space-occupying lesion affecting the basal ganglia
  - post encephalitis
  - Wilson's disease
  - 'Parkinson's plus' syndromes: progressive supranuclear palsy (Steele–Richardson–Olszewski's syndrome), multisystem atrophy (Shy–Drager's syndrome)

— other disorders resembling parkinsonism include dementia (including normal-pressure hydrocephalus) and a benign essential tremor.

Investigations

- Parkinson's disease is generally a clinical diagnosis
- Supportive evidence may be provided by a therapeutic trial of levodopa or an apomorphine challenge
- Cerebral imaging may be appropriate where other diagnoses require exclusion
- A dopamine transporter (DaT) scan can be used to distinguish Parkinson's disease from a benign tremor

Management

- MDT approach
  — neurologist, physiotherapist, OT, specialist nurse, SALT
- Levodopa
  — traditionally the pivotal treatment; convention now suggests it should be commenced when the Parkinson's disease is having a significant impact upon lifestyle, as it is effective for only a limited number of years (after this time, patients often experience fluctuation between dyskinesia and immobility; the 'on–off' effect)
- Other medications
  — dopamine agonists
  — anticholinergics
  — selegiline
  — entacapone
  — apomorphine
- Complications
  — assess disability and cognition regularly; treat depression
- Surgery
  — e.g. deep brain stimulation;[2] may be considered for those inadequately controlled by drugs.

Questions

1 'What are the characteristic features of Parkinson's disease?'
  — The triad of parkinsonism. In Parkinson's disease these features begin asymmetrically and generally affect the upper limbs first. There is fluctuation in severity, with the patient being better some days than others.

- Autonomic dysfunction and changes in higher mental functioning are also common.

2 'What drugs other than L-dopa are used in the management of Parkinson's disease?'
- Dopamine agonists (e.g. ropinirole): may be used as first-line therapy, especially in younger patients (due to the decreased risk of dyskinesias), or as an add-on to levodopa.
- Anticholinergics (e.g. trihexyphenidyl): useful for tremor.
- Monoamine oxidase-B inhibitors (selegiline): may be helpful with motor symptoms.
- COMT inhibitors (entacapone): help to decrease immobility by shortening the 'off' time associated with L-dopa.
- Apomorphine: a parenteral dopamine agonist that can be helpful with 'on–off' effects. It may be given via a syringe driver.

3 'What is the pathology behind Parkinson's disease?'
- Degeneration of the substantia nigra dopaminergic neurones in the basal ganglia. The hallmark is the presence of Lewy bodies.

Key points

- Have a specific examination prepared for Parkinson's disease.
- Look for the triad of parkinsonism early, and then search for extra features.

Peripheral neuropathy

'Please examine this patient who has painful legs.'

Findings

- General
  - walking aids, diabetic shoes, prostheses, insulin pen
- Neurological
  - 'glove and stocking' sensation loss
  - all sensory modalities should be lost (use a 128 Hz tuning fork to assess vibration sense as dorsal columns are usually affected first)
  - wasting, weakness, areflexia
- Extras
  - finger prick testing, cataracts, ulcers, Charcot's joints, callus (diabetes mellitus)
  - clawing of the toes, pes cavus (Charcot–Marie–Tooth's disease)
  - amiodarone facies
  - anaemia (B12 deficiency)
  - evidence of alcohol abuse
  - evidence of arthritis and rashes (vasculitis)

Presentation

'This patient has a predominantly sensory peripheral neuropathy as evidenced by lack of sensation bilaterally to mid-calf for all modalities. There is no evidence of any ulceration or callus formation. The most likely underlying cause is diabetes mellitus as evidenced by finger pulp pricks from capillary blood glucose testing.'

- It is important to first ensure that the patient has a peripheral neuropathy rather than a mononeuropathy or mononeuritis multiplex.
- This is a common PACES case and looking for a cause is crucial. Try to decide if the neuropathy is predominantly motor or sensory as this should help to elucidate the aetiology.
- Predominantly sensory:
  - diabetes mellitus
  - alcohol

  - drugs
  - vitamin deficiencies (B1, B12)
  - uraemia.
- Predominantly motor:
  - Guillain–Barré syndrome
  - malignancy
  - Charcot–Marie–Tooth's disease
  - porphyria
  - lead poisoning.
- Other causes include paraneoplastic syndromes, paraproteinaemia, vasculitis and infections (HIV, Lyme disease). Some cases are idiopathic.

Investigations

- Full drug and alcohol history
- Blood tests: FBC (including MCV), U&Es, LFTs (including GGT), vitamin B12 and folate, glucose, TFTs, autoimmune screen and immunoglobulins, hepatitis screen, HIV screen, Lyme serology
- Urine: dip for glucose and protein, Bence Jones's protein
- Imaging: CXR
- LP and CSF study: protein and CSF virology
- Nerve conduction studies

Management

- Dependent upon the aetiology; remove any precipitants and treat the cause

Questions

1 'What is autonomic neuropathy?'
  - A neuropathy of the autonomic nervous system.
  - May present alone or in conjunction with a motor or sensory neuropathy; the commonest cause is diabetes.
  - May present with postural hypotension, impotence, urinary retention, diarrhoea/constipation and a Horner's syndrome.

5 'What is Charcot–Marie–Tooth's disease?'
  - A hereditary sensory and motor neuropathy. Also known as peroneal muscular atrophy.
  - Usually starts at puberty with foot drop and weak legs.
  - The peroneal muscles are the first to atrophy, with upper limb signs appearing at a later stage.

  - There is muscle wasting, pes cavus and a bilateral foot drop (high-stepping gait). Reflexes are often absent. Sensory loss is variable.
  - The most common form is inherited in an autosomal dominant manner.
6. 'Which drugs can cause a peripheral neuropathy?'
  - Amiodarone
  - Gold
  - Isoniazid
  - Metronidazole
  - Nitrofurantoin
  - Phenytoin
  - Vinca alkaloids

Key points

- Establish the diagnosis early to allow a cause to be found.
- Be sure to present a list of possible aetiologies to the examiner, taking into account whether the neuropathy is predominantly motor or sensory.
- The most likely causes in PACES include diabetes and Charcot–Marie–Tooth's disease.

Mononeuropathies

'Please examine this patient who has paraesthesiae in his left hand.'

Findings

Carpal tunnel syndrome

- Wasting of the thenar eminence
- Weakness of LOAF (lumbricals, opponens pollicis, abductor pollicis brevis, flexor pollicis brevis)
- Sensory loss over the lateral 3½ digits
- Maximal wrist flexion for one minute may elicit symptoms (Phalen's test)
- Tapping over the nerve at the wrist induces tingling (Tinel's test)
- Look for a scar from carpal tunnel release surgery and evidence of diabetes, hypothyroidism, acromegaly and rheumatoid arthritis

Presentation

'This patient has wasting of the thenar eminence in the left hand, indicating a median nerve lesion. This is supported by weakness of opposition of the thumb. There is sensory loss over the lateral 3½ digits with sparing of the palm. Tinel's test is positive. There is a scar over the anterior wrist. The patient also has a symmetrical arthritis affecting the hands with a swan-neck deformity of two digits, but no evident skin or nail changes. The likely diagnosis is carpal tunnel syndrome of the left wrist on a background of rheumatoid arthritis.'

Ulnar nerve palsy

- Wasting of the hypothenar eminence (thenar eminence spared); claw hand; guttering on the dorsal aspect of the hand
- Also look for wasting of the medial aspect of the forearm (note low/high lesions; hand less clawed in high lesion)
- Weakness of abduction and adduction (test for Froment's sign) of the fingers and adduction of the thumb
- Sensory loss over the medial 1½ digits
- Look for scars (fracture dislocation) and osteoarthrosis at the elbow

Radial nerve palsy

- Wrist drop; weakness of wrist extension; if the wrist is passively extended, intrinsic muscles of the hand should be intact
- Impaired grip strength
- Sensory loss over the first dorsal interosseous
- Look for scars at the elbow (fracture/dislocation) and note if the patient uses crutches

Common peroneal nerve palsy

- Foot drop on inspection (leading to a high-stepping gait)
- Weakness of dorsiflexion and eversion of the foot
- All reflexes will be intact
- Sensory loss over the lateral dorsum of the foot
- Look for evidence of compression around the fibular neck
- In all mononeuropathies it is important to look for causes of mononeuritis multiplex

Presentation

- Functionality is crucial and should always be tested and commented upon; try to find a cause for the lesion

Investigations

- Single nerve lesions are often a clinical diagnosis; neurophysiology can be used to confirm the diagnosis and to assess severity

Management

- Splints and physiotherapy can be helpful; surgery is sometimes used, especially in carpal tunnel syndrome where surgical decompression of the flexor retinaculum is a simple and definitive treatment

Questions

1 'What is the differential diagnosis of a foot drop?'
   — Common peroneal nerve palsy, peripheral neuropathy (especially Charcot–Marie–Tooth's disease), sciatic nerve palsy, L4/5 radiculopathy (prolapsed lumbar disc) and lumbosacral plexopathy.

2 'What are the causes of mononeuritis multiplex?'
   — Wegener's granulomatosis, rheumatoid arthritis, diabetes mellitus, polyarteritis nodosa, sarcoidosis, amyloidosis, carcinomatosis, leprosy.

3 'What are the causes of wasting of the small (intrinsic) muscles of the hand?'
   — Resembles an ulnar nerve lesion, but with thenar wasting and weakness also.
   — Causes include:
      - anterior horn cells disease, e.g. poliomyelitis
      - radiculopathy, e.g. trauma, prolapsed disc
      - plexopathy, e.g. brachial plexus injury, Pancoast's tumour, cervical rib
      - peripheral nerve lesions
      - muscle, e.g. disuse atrophy.

Key points

- Initially look for any evident wasting or deformity.
- Examine all motor and sensory areas supplied by the nerve to localise the lesion.
- Search for a possible cause.

Motor neurone disease

'Please examine this patient who presents with weakness.'

Findings

ALS

- Both UMN and LMNs affected so maybe flaccid or spastic
- Weakness throughout; degree of weakness depends on the number of muscles affected and distribution of motor neurone loss
- Reflexes usually exaggerated (UMN and LMN signs)
- Ankle clonus elicited
- Bilateral extensor plantar responses
- Sensation is unaffected throughout
- Involvement of lower cranial nerves causes a pseudobulbar palsy

PMA

- Flaccid weakness as only LMNs affected
- Fasciculations and wasting
- Decreased or absent reflexes
- Plantars down-going

PLS

- UMN signs only
- Usually begins in lower limbs (spastic gait)
- Exaggerated reflexes

Progressive bulbar palsy

- Only lower cranial nerves affected (IX, X, XII)
- 'Donald Duck'/nasal speech
- Weakness of palatal muscles results in swallowing difficulties

Presentation

'The most likely diagnosis is motor neurone disease. There is generalised wasting and fasciculation. Tone is increased with generalised weakness. Reflexes are brisk/reduced/absent. Leg reflexes are brisk and ankle clonus is present. There are bilateral extensor plantar responses.'

Investigations

- Diagnosis is mainly clinical, based on a high index of suspicion from the collection of signs and symptoms
- EMG
  - abnormally slowed conduction due to reduction in the number of viable motor axons/anterior horn cells to activate the muscle(s) involved
- Nerve conduction studies
  - normal sensory nerve conduction and abnormal motor nerve conduction
  - reduced muscle action potentials
  - repetitive stimulation: decremental response with slow repetitive stimulation
- MRI
  - to exclude other causes for the symptoms, e.g. cervical myelopathy/spondylosis or cord compression

Management

- Supportive measures
  - physiotherapy, occupational therapy, speech therapy
  - swallowing and nutritional support (NG/PEG feeding)
  - respiratory support: NIPPV, tracheostomy and invasive ventilation
- Specific therapy
  - riluzole: a glutamate inhibitor that acts by inhibiting voltage-gated sodium channels
  - increases survival compared to placebo by roughly three months; no significant effect on muscle strength or neurological function seen

Questions

1 'What are the disease variants of MND?'
  - Amyotrophic lateral sclerosis (ALS)
    - ~50% of cases; combined UMN and LMN signs; mostly sporadic (90%–95%)
    - familial form: copper/zinc superoxide dismutase (SOD-1) gene mutation on chromosome 21
  - Primary lateral sclerosis (PLS)
    - rare; affects upper motor neurones only; has the best prognosis but can later progress to ALS

- Progressive muscular atrophy (PMA)
  - ~25% cases; affects anterior horn cells only therefore signs in distal muscle groups
- Progressive bulbar palsy
  - ~25% cases; worst prognosis; affects suprabulbar nuclei and lower cranial nerves resulting in speech and swallowing difficulties (increased risk of aspiration)

2 'What is the prognosis for patients with MND?'
- No known cure; usually fatal within 3–5 years of diagnosis; cause of death is usually aspiration pneumonia and/or ventilatory failure

3 'What is the differential diagnosis for MND?'
- Degenerative: cervical cord compression, cervical spondylosis
- Inflammatory/traumatic/inherited: syringomyelia, spinal muscular atrophy
- Infectious: polio, syphilis
- Malignant/paraneoplastic

4 'What are the other causes of absent ankle jerks and extensor plantar responses?'
- Hereditary cerebellar ataxias: Friedreich's 's ataxia, spinocerebellar ataxia
- Syphilitic taboparesis
- Subacute combined degeneration of the cord
- Conus medullaris pathology
- Combined pathologies, e.g. peripheral neuropathy from any cause in a patient with cervical spondylosis

Key points

- Note any external features that may indicate the diagnosis (fasciculations, NIPPV machine).
- Note the patient's speech.
- It is crucial to check the plantar response.

Multiple sclerosis

'Please examine the legs of this patient who has progressive difficulty in walking.'

Findings

- General
  - walking sticks, wheelchair, catheter, ataxic gait, dysarthria, mood (depressed/elated)
- Eyes
  - internuclear ophthalmoplegia, optic neuritis, central scotoma, loss of colour vision, relative afferent pupillary defect (RAPD), nystagmus
- Neurological
  - spastic paraparesis, cerebellar signs

Presentation

'This patient has an ataxic gait. Examination of her lower limbs shows increased tone and brisk reflexes. Further examination shows that the patient also has evidence of bilateral internuclear ophthalmoplegia. Putting these findings together, this patient is likely to have a diagnosis of multiple sclerosis.'

- Comment on the patient's functionality by looking for general findings, e.g. walking aids, the presence of a catheter and the patient's general affect.

Investigations

- CSF analysis
  - IgG oligoclonal bands on electrophoresis
- Visual-evoked potentials
  - delayed response
- MRI
  - highlighting areas of demyelination

Management

- Multidisciplinary approach/patient education/MS support group information

- Acute relapse
  - IV methylprednisolone: this may help to reduce the duration and severity of the relapse but will not alter the course of the disease
- Disease-modifying drugs
  - interferon beta-1a
  - interferon beta-1b
  - glatiramer acetate
  - azathioprine
  - natalizumab (for highly active relapsing–remitting disease)
- Symptomatic treatment
  - spasticity: physiotherapy, baclofen, tizanidine
  - urinary dysfunction: oxybutynin, catheterisation
  - constipation: laxatives, enemas
  - pain: amitriptyline, carbamazepine, gabapentin
  - fatigue: amantadine
  - depression: support groups, SSRIs

Questions

1 'What is multiple sclerosis?'
  - Multiple sclerosis is a chronic inflammatory autoimmune disease of the central nervous system. The diagnosis depends on demonstrating at least two demyelinating lesions in the brain or spinal cord on MRI, separated in time and space.

2 'What are Lhermitte's sign and Uhthoff's phenomenon?
  - Lhermitte's sign
    - flexion of the neck causing an 'electric shock'-like sensation in the trunk and limbs; this occurs in cervical spondylosis as well as MS
  - Uhthoff's phenomenon
    - an increase in the severity of symptoms (mainly visual), commonly precipitated by an increase in temperature or by exercise

3 'What subtypes are used to classify multiple sclerosis?'
  - Relapsing–remitting
    - affects approximately 85% of MS sufferers
  - Secondary progressive
    - follows a period of relapsing–remitting MS
  - Primary progressive
    - 15% of cases; progressive deterioration from the start

— The classification of MS becomes important when considering the role of DMARDs. Currently these drugs can help reduce the severity and frequency of relapses in relapsing–remitting MS and secondary progressive MS.

Key points

- A combination of spastic paraparesis and cerebellar signs is likely evidence of multiple sclerosis.
- Always comment on the patient's functional status.

Spastic paraparesis

'Please examine this patient's legs neurologically as they are finding it difficult to walk.'

Findings

- General
  - walking aids, scissoring gait
- Neurological
  - increased tone, clonus, decreased power, hyperreflexia, extensor plantars, and sensory loss; signs may be present only below a particular spinal level
- Other
  - evidence of cerebellar syndrome
  - cachexia and other evidence of malignancy

Presentation

'On examination this patient has signs suggestive of a spastic paraparesis. This is evidenced by hypertonia, hyperreflexia and decreased power throughout the lower limbs. There is tenderness over the T12 vertebra with a sensory level present, making spinal cord compression a likely cause.'

- Comment on a possible underlying cause for the spastic paraparesis
  - Compression
    - tumour, osteoarthritis, trauma/fracture, central disc prolapse
  - Transverse myelitis
    - multiple sclerosis (cerebellar signs), inflammatory and vascular disorders
  - Degenerative
    - hereditary spastic paraparesis, motor neurone disease (absence of sensory signs/combination of UMN and LMN signs), Friedreich's ataxia (cerebellar signs)
  - Infective
    - HIV myelopathy
  - Others
    - cerebral palsy, subacute combined degeneration of the cord

- It is important to exclude acute spinal cord compression, so tell the examiners you would ask about bladder and bowel symptoms, and would offer to perform a PR examination.
- Always check gait and assess the functional status of the patient.

Investigations

- MRI of the spine is the gold-standard imaging test
- Other investigations include: FBC, vitamin B12/folate levels, ESR/CRP, syphilis serology, CSF protein and oligoclonal bands, myeloma screen and tumour markers

Management

- The most urgent treatment is needed for spinal cord compression:
  - if malignancy is the cause, give dexamethasone; this is followed by radiotherapy or surgery
  - surgery is the mainstay of treatment for other causes
- Further management of this presentation depends upon the underlying cause, though an MDT approach focusing on neurorehabilitation may be helpful

Questions

1 'What malignancies are most likely to cause spinal cord compression?'
  - The most likely cause is metastasis from lung, breast or prostate cancer. Kidney and thyroid primaries also commonly metastasise to bone.
  - Multiple myeloma is also a relatively common cause of cord compression.
  - Intrinsic spinal malignancy is rare.

2 'What is transverse myelitis?'
  - Inflammation of the spinal cord, characterised by axonal demyelination. Generally the inflammation is across the thickness of the cord.
  - Symptoms come on over a period of hours to weeks.
  - The most common cause is multiple sclerosis. Viral and other infections can be implicated. Often no cause is found.
  - Steroids and plasma exchange may be helpful in management, though more crucial is neurorehabilitation.

Key points

- When finding a spastic paraparesis it is important to look for a spinal level.
- Be sure to assess gait and examine the spine locally.
- Excluding acute cord compression is essential.

Visual field defects

'Please examine this patient who has recently developed a visual disturbance and as a result, finds that they are bumping into things.'

Tips

- Sit at the same height as the patient. Ask the patient to cover each eye in turn.
- Assess both nasal and temporal fields and all four quadrants. Look carefully at where the field of vision starts/finishes and which areas of the field are lost, i.e. nasal/temporal, superior/inferior.
- Use the information collected to establish which pattern of visual field defect the patient has.
- Be sure you are familiar with the visual pathway anatomy to enable you to ascertain the site of the lesion, as the nature of the visual field defect is determined by the site of the lesion along the visual pathway.

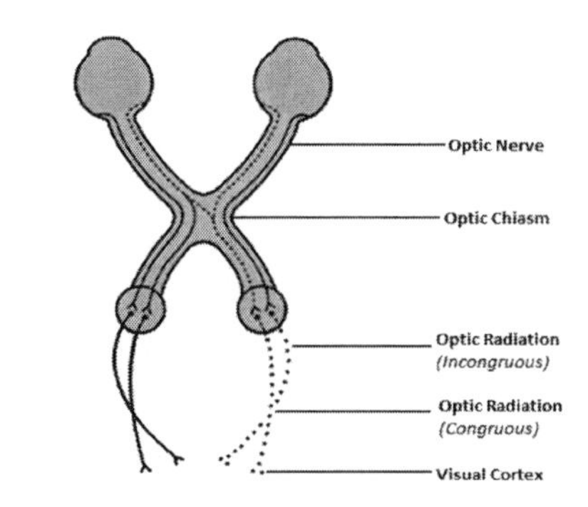

Figure 3.1 Visual pathway

Presentation

'This patient presents with a history of bumping into objects and visual impairment. They have visual loss affecting the same half of the visual field in each eye. This is known as a homonymous hemianopia.'

- Classification of defects depends on whether one or both eyes are affected, which half of the visual field is affected, i.e. temporal, nasal or both (heteronymous or homonymous hemianopia), and by the extent of field affected in each eye relative to the other (congruity). *See* Figure 3.1.

- Optic nerve lesions: result in partial or complete visual loss on the side of the lesion.
- Optic chiasm lesions: result in both temporal fields being lost.
- Optic radiation lesions: result in homonymous field defects that depend on the location of the lesion in the temporal or parietal lobe.
- *Temporal* lobe lesions lead to *superior* homonymous quadrantanopias.

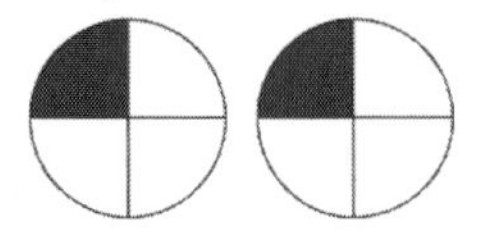

- *Parietal* lobe lesions lead to *inferior* homonymous quadrantanopias.

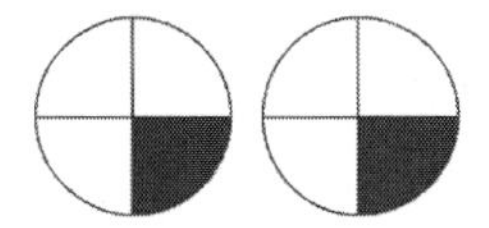

- The further back towards the visual cortex the defect, the greater the degree of congruity (i.e. both eyes affected to the same degree).

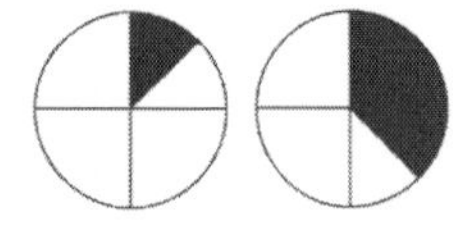

Incongruous defect

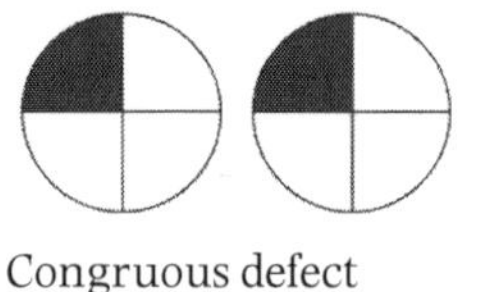

Congruous defect

- Visual cortex or optic radiation lesions result in a homonymous hemianopia.

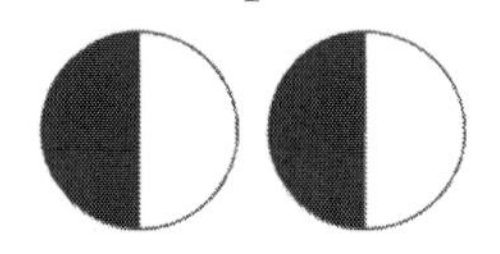

Questions

1 'What are the causes of a homonymous hemianopia?'
- Vascular: stroke
- Trauma
- Tumour
- Infection: encephalitis
- Demyelination: MS

2 'What are the causes of an optic nerve lesion?'
  - Trauma
  - Inflammatory: demyelination, optic neuritis
  - Compression: raised intraocular or intracranial pressure
  - Vascular: acute optic artery ischaemia
  - Metabolic: B12 deficiency, diabetes, alcohol excess
  - Inherited conditions, e.g. Leber's optic neuropathy

3 'What are the common causes of a bitemporal hemianopia?'
  - Usually occur as a result of a lesion of the optic chiasm
    - pituitary fossa tumour
    - craniopharyngioma
    - suprasellar meningioma

Key points

- Remember to assess both temporal and nasal visual fields for both eyes.
- Be familiar with visual pathways so the site of the lesion can be identified.

Neurology station summary

- Neurology is a difficult station, particularly if you don't look well practised. You must practise your examinations until you really are slick.
- As ever, particular emphasis should be paid to observation. In neurology many diagnoses can be made from the end of the bed (Parkinson's/myasthenia gravis/myotonic dystrophy); when this is the case don't relax! You still need to demonstrate all the signs to the examiner.
- Looking around the bed is just as important in neurology as any other station (e.g. look for FEV_1 monitors in myasthenia gravis).
- Don't panic if you don't know the diagnosis; work out where the lesion is and use a pathological sieve to create a differential list.
- Be aware of how to demonstrate specific signs for specific syndromes (e.g. myotonia in myotonic dystrophy); your examiners will be impressed if you look slick whilst doing this.
- Go into the station prepared to do a range of examinations. You may have to examine the arms, the legs, the eyes, the cranial nerves or a special examination.
- Don't forget to reinforce the reflexes if you think that they are absent.
- Don't forget to check sensation on the chest prior to commencing limb examination, so that comparison can be made.
- Asking the patient to walk so that you can assess their gait is a key skill in neurology. Be aware of the different types of gait and the different syndromes that each gait may indicate.
- When presenting your case first tell the examiners the syndrome or the likely site of the lesion, followed by the likely cause.

References

1 Bamford J, Sandercock P, Dennis M, *et al*. Classification and natural history of clinically identifiable subtypes of cerebral infarction. *Lancet*. 1991; **337**: 1521–6.
2 Benabid AL; Chabardes S; Mitrofanis J, *et al*. Deep brain stimulation of the subthalamic nucleus for the treatment of Parkinson's disease. *Lancet Neurol*. 2009; **8**(1): 67–81.

Station 3: Cardiology

Contents

Hints for the cardiology station

- Pick up on peripheral signs as these will help guide you to a diagnosis by the time you reach for your stethoscope!
- Check the radial pulse; comment on rate, rhythm or absence.
- Look at the carotid pulsation from the end of the bed; a 'dancing carotid' signifies Corrigan's sign.
- Examine the JVP with the patient resting at a 45-degree angle; use the hepatojugular reflex if not visible.
- Giant 'V' waves in the JVP may indicate tricuspid regurgitation.
- Feel for a parasternal heave (indicative of right ventricular hypertrophy).
- Check for a mitral valvotomy scar under the left breast; this can be easily missed.
- Always comment on the position of the apex beat and demonstrate the position by marking it out on examination and re-establish its location in the left lateral position before auscultation.
- If you cannot feel the apex beat in the 5th intercostal space/mid-clavicular line, the patient may have evidence of cardiomegaly or dextrocardia.
- Listen from the end of the bed for an audible click from metallic heart valves, and comment on any evidence of valvular regurgitation or endocarditis.
- On auscultation there may be more than one murmur; try to identify the dominant one.
- Ensure you expose the legs up to the thighs at the end of the examination to look for any scars from vein grafts or evidence of peripheral vascular disease.
- Comment on any evidence of infective endocarditis, heart failure and other significant comorbidities.

Aortic stenosis

'This patient has had episodes of collapse and chest pain. Please examine their cardiovascular system.'

Findings

- General
  - dyspnoea
- Peripheral
  - slow-rising pulse, low systolic blood pressure, narrow pulse pressure
- Chest
  - palpable thrill over the aortic area, heaving apex beat, ejection systolic murmur loudest over the aortic region radiating to the carotids, soft or absent A2, reversed splitting of S2, S4

Presentation

'This patient has aortic stenosis as evidenced by a slow-rising pulse, narrow pulse pressure, heaving apex beat and an ejection systolic murmur that radiates to the carotids. I believe in this case it is a severe stenosis as the aortic 2nd heart sound is inaudible and I can hear a 4th heart sound. There is no evidence of congestive cardiac failure.'

- Look for evidence of congestive cardiac failure, a displaced apex, associated mitral regurgitation (or Gallavardin's phenomenon due to radiation of the murmur through the LA to the apex mimicking MR), and signs of pulmonary hypertension (left parasternal heave, loud A2).
- Try to comment on the severity of the murmur.

Differential diagnosis

- Aortic sclerosis (normal A2)
- HCM (look for jerky pulse, double apical impulse, thrill and ejection systolic murmur loudest at left lower sternal edge)
- Flow murmur (pulmonic on inspiration, especially in young women)
- Pulmonary stenosis

Investigations

- Diagnosis
  - ECG (large LA, voltage LVH, LBBB)
  - chest X-ray (may be normal, unfolded aorta)
  - echocardiogram (investigation of choice) ± TOE (severity of AS, concomitant valvular lesions, LV dimensions and function)
  - cardiac catheterisation (underlying coronary artery disease)
  - cardiac MRI scan can assess severity of AS and myocardial viability

Management

- Medical
  - regular follow-up and echocardiograms (timing for intervention)
  - diuretic therapy for heart failure
- Surgical
  - aortic valve replacement (AVR) +/– CABG (mainstay of treatment of symptomatic severe aortic stenosis)
  - transcatheter aortic valve implantation (TAVI) considered if open AVR too high risk (transfemoral or apical)

Questions

1. 'What are the causes of aortic stenosis?'
   - Common
     - senile degenerative calcification
     - bicuspid valve
   - Rare
     - rheumatic fever
     - HOCM
     - congenital (other than bicuspid): supravalvular stenosis (Williams's syndrome)
2. 'What are the indications for surgery?'
   - Severe stenosis (valve area $<1.0\,cm^2$ on echo, peak velocity $>4\,m/s$) **or**
   - Symptoms (angina/collapse/dyspnoea/heart failure) with moderate stenosis (valve area $<1.5\,cm^2$ on echo, peak velocity $>3\,m/s$)
3. 'What differentiates severe aortic stenosis from aortic sclerosis?'
   - Difficult! They lie on a continuum. In sclerosis there is a normal pulse pressure and character, normal A2 component and little murmur radiation.

4 'What happens to the loudness of the murmur with progressive stenosis severity?'
   — Murmur intensity is dependent on the flow turbulence through the valve and the cardiac output. Thus in critical AS with a failing ventricle, cardiac output will fall and the murmur will be soft, but A2 will be absent.

Key points

- Aortic stenosis is an ejection systolic murmur which is loudest over the aortic area and radiates to the carotids.
- Comment on the peripheral signs associated with this murmur.
- Ensure that you know the indications for surgical treatment.

Aortic regurgitation

'This patient complains of progressive breathlessness. Examine the patient's cardiovascular system.'

Findings

- General
  - dyspnoea
- Peripheral
  - collapsing (waterhammer) pulse, wide pulse pressure
- Chest
  - displaced and hyperdynamic apex beat, high-pitched early-diastolic murmur (heard best in fixed expiration at the left sternal edge or sometimes the right sternal edge)
- Eponymous signs
  - Corrigan's sign: 'dancing carotid'
  - De Musset's sign: head nodding
  - Quincke's sign: pulsation at nail-bed when pressed
  - Traube's sign: pistol-shot sound heard over the femoral arteries
  - Austin Flint's murmur: mitral stenosis murmur due to impingement of the anterior mitral valve leaflet by the regurgitant jet

Presentation

'This patient has aortic regurgitation as evidenced by the early-diastolic murmur which is heard loudest at the left sternal edge with the patient sitting forward. This patient also has a collapsing pulse and evidence of Corrigan's sign.'

- Comment on the severity of the aortic incompetence. In severe cases there may be a widened pulse pressure, soft S2, short EDM, S3 sound and evidence of left ventricular failure.
- Comment on the possible underlying cause of AR
  - Acute
    - infective endocarditis
    - aortic dissection

- Chronic
  - congenital aortic valve malformation (e.g. bicuspid, quadricuspid)
  - aortic root dilatation (annulo-aortic ectasia)
  - prior endocarditis
  - rheumatic fever
- Connective tissue disease
  - rheumatoid arthritis, SLE, ankylosing spondylitis, Ehlers–Danlos syndrome, osteogenesis imperfecta
- Marfan's syndrome
  - high-arched palate, tall, arachnodactyly
  - autosomal dominant condition due to mutations of protein fibrillin 1 located on chromosome 15
- Seronegative arthritides
  - ankylosing spondylitis, Reiter's syndrome
- Syphilitic aortitis
  - rare cause, may be associated with Argyll Robertson's pupils

Investigations

- CXR
  - cardiomegaly ('coeur en sabot'), evidence of LVF
- ECG
  - lateral T-wave inversion, left ventricular hypertrophy
- ECHO ± TOE
  - to determine if aortic regurgitation is due to valve or root disease and to assess left ventricular size and systolic function
- CT or MRI
  - of aorta if aortic root disease

Management

In acute cases (infective endocarditis and aortic dissection) urgent surgery may be needed.

In chronic cases, regular monitoring of the patient by a cardiologist to determine when a valve replacement is appropriate depending on the patient's symptoms and clinical findings.

- Medical
  - Regular follow-up and echocardiograms (timing for intervention)
  - Diuretic therapy and ACE inhibitors for heart failure

- Surgical
  - Aortic valve replacement (AVR) ± CABG
  - transcatheter aortic valve implantation (TAVI) considered if open AVR too high risk (transfemoral or apical)

Questions

1 'List the findings which would determine the need for surgery in aortic regurgitation.'
  - Symptomatic patients with severe AR (dyspnoea, NYHA II-IV, angina)
  - Asymptomatic patients
    - left ventricular ejection fraction of ≤50%
    - left ventricular ejection fraction of >50% and end-systolic dimension of >50 mm (aortic valve)

2 'List some other manifestations of Marfan's syndrome.'
  - Ectopia lentis (upwards lens dislocation), arm span > height, dural ectasia, pectus excavatum, joint laxity, scoliosis, pes planus (Ghent criteria).

3 'What is an Austin Flint's murmur?'
  - This is a low-pitched, rumbling mid-diastolic murmur which is a sign of severe aortic regurgitation. It is attributed to the fluttering of the anterior mitral valve leaflet caused by a severe regurgitant stream.

Key points

- Aortic regurgitation is an early-diastolic murmur which is heard loudest over the left sternal edge with the patient sitting forward.
- Ensure that you can recognise the eponymous signs associated with this murmur.
- Know the common acute and chronic causes of this murmur.

Eisenmenger's syndrome

'Please examine this patient who presents with dyspnoea and orthopnoea.'

Findings

- General
  - — cyanosis
- Peripheral
  - — clubbing, raised JVP
- Chest
  - — left parasternal heave
  - — loud single P2 audible (and palpable)
  - — possibly features of a VSD, an ejection systolic click in the pulmonary area, an audible early-diastolic murmur in the pulmonary area due to pulmonary regurgitation, and/or a pansystolic murmur at the LSE due to tricuspid regurgitation (secondary to pulmonary hypertension)
  - — possible wide fixed-split first heart sound (ASD)

Presentation

'This patient has features suggestive of Eisenmenger's syndrome. They are peripherally cyanosed with clubbing. There is a left parasternal heave. On auscultation there is a loud pulmonary second heart sound with an audible ejection click in the pulmonary area.'

Investigations

- ECG
  - — signs of right ventricular hypertrophy with P-pulmonale
- Echocardiogram
  - — to assess the cause and degree of shunt (VSD/PDA), the presence of valvular disease (TR/PR), and to calculate pulmonary artery pressures and the extent of pulmonary hypertension
- CXR
  - — increased pulmonary vascular markings, prominent right ventricle, large RA

- Cardiac catheterisation
  - to assess the degree of shunting, and to calculate pressure gradients and pulmonary pressures

Management

- Ideally, treat defects early to prevent this syndrome from developing
- Medical
  - oxygen: symptomatic benefit (no prognostic benefit)
  - vasodilators: prostacyclin, bosentan, sildenafil
  - diuretics to treat right heart failure
  - contraceptive measures strongly advised: pregnancy carries high maternal and foetal mortality rates
- Surgical
  - not beneficial once Eisenmenger's is present, due to high mortality
  - palliative procedures are available
  - combined heart-lung transplantation is an option

Questions

1 'What is Eisenmenger's syndrome?'
  - Eisenmenger's results from a large left-to-right shunt/cardiac defect causing increased pulmonary blood flow and resultant pulmonary hypertension. This leads to reversal of the shunt causing either a unidirectional right-to-left shunt or a bidirectional shunt. This is clinically manifest as cyanotic heart disease.

2 'What are the causes of Eisenmenger's syndrome?'
  - Large non-restrictive VSD
  - Non-restrictive PDA
  - Atrioventricular septal defects
  - Large uncorrected or surgically created systemic-to-pulmonary shunts for treatment of congenital heart disease

Key points

- Ensure that you comment on the presence of cyanosis and clubbing.
- Be aware of the underlying causes of this condition.
- Be aware that once Eisenmenger's develops, treatment of the underlying defect is purely medical.

Hypertrophic cardiomyopathy

'This patient, who has had episodes of syncope and palpitations, has been told he has a murmur. Please examine his cardiovascular system.'

Findings

- General
  - dyspnoea
- Peripheral
  - bifid jerky pulse, large 'A' wave visible in JVP, low systolic blood pressure, narrow pulse pressure
- Chest
  - palpable thrill over the aortic area, heaving double apical impulse, ejection systolic murmur loudest over the left sternal edge (accentuated by Valsalva manoeuvre, softer with squatting), S4, normal A2
- Extra
  - look for evidence of congestive cardiac failure, murmur of MR

Presentation

'This patient probably has HCM as evidenced by a jerky pulse, double-impulse apex beat and an ejection systolic murmur that radiates to the carotids but a normal A2 component. There is no evidence of congestive cardiac failure.'

- Mention the murmur and comment on severity. Comment on lack/presence of signs of congestive cardiac failure.

Differential diagnosis

- Aortic stenosis
- Flow murmur
- Pulmonary stenosis

Investigations

- ECG
  - voltage LVH, deep Q waves, interventricular conduction delays
- Chest X-ray

- Echocardiogram
  - investigation of choice, looking for septal asymmetrical hypertrophy, systolic anterior motion (SAM) of the mitral valve, MR, diastolic LV dysfunction
- Cardiac MRI
  - demonstrates severity and distribution of hypertrophy and wall fibrosis
- Holter monitor
  - non-sustained VT
- Exercise tolerance test
  - blood pressure response, looking for inadequate rise or paradoxical systolic drop with exertion
- Cardiac catheterisation
  - aortic outflow gradient, coronary artery disease

Management

- Medical
  - regular follow-up and echocardiograms
  - beta-blocker/calcium-channel antagonist (verapamil) ± disopyramide
  - diuretic therapy only for heart failure
  - septal alcohol ablation
  - permanent pacemaker insertion
  - implantable cardioverter defibrillator (ICD) insertion
- Surgical
  - septal myomectomy
  - heart transplant considered for severe refractory heart failure

Questions

1 'What is the cause of the dynamic outflow gradient?'
  - It is caused by systolic anterior motion of the mitral valve due to local under-pressure (Venturi effect), exacerbated by the septal hypertrophy.

2 'What are the risk factors for sudden cardiac death (SCD)?'
  - Family history of SCD with HCM, syncope, young age (<30 years), sustained VT, septal thickness >30 mm, hypotensive response to exercise.

3 'What do you know about the genetics of HCM?'

— HCM is inherited as an autosomal dominant trait or arises spontaneously producing a de novo mutation (usually of the β-myosin heavy-chain gene).

— The majority of cases are due to mutations in the β-myosin heavy-chain gene (chromosome 14) and cardiac myosin-binding protein C gene (chromosome 11) but also cardiac troponin I and α-tropomysin genes.

— Therefore a detailed family history and first-degree relative screening should be advised.

Key points

- HCM is associated with an ejection systolic murmur heard loudest at the left sternal edge.
- Key clinical findings are a bifid jerky pulse, double apical impulse and a palpable thrill.
- This is an autosomal dominant, inherited condition which is associated with sudden cardiac death.

Mitral stenosis

'This patient complains of breathlessness. Please examine their cardiovascular system.'

Findings

- General
  - malar flush
- Peripheral
  - irregularly irregular pulse, raised JVP
- Chest
  - inspection/palpation: left thoracotomy/valvotomy scar, tapping apex, palpable S1
- Auscultation
  - opening snap of mitral valve, loud S1, low-pitched rumbling mid-diastolic murmur at the apex (heard loudest in the left lateral position)

Presentation

'This lady has a malar flush. There is an irregularly irregular pulse but no stigmata of endocarditis. The apex beat is not displaced. There is a loud first heart sound and there is a mid-diastolic murmur heard loudest in the left lateral position. This is mitral stenosis.'

- Look out for signs of pulmonary hypertension (left parasternal heave, loud/palpable P2).

Investigations

- Diagnosis
  - chest X-ray: left atrial enlargement
  - echocardiogram ± TOE: to assess the mitral and other valves, and left atrial dimension
- Complications
  - ECG: AF, right ventricular hypertrophy
  - echocardiogram: to assess for pulmonary hypertension and evidence of endocarditis
  - cardiac catheterisation
  - blood cultures: if endocarditis suspected

Management

- Medical
  - diuretics
  - treat AF with rate control
  - anticoagulate all patients with warfarin where possible
- Surgical
  - indications include significant symptoms which limit normal activity, recurrent emboli, pulmonary oedema (especially in pregnancy) and deterioration due to AF
  - options include mitral valve replacement, open valvotomy and balloon valvuloplasty

Questions

1 'What are the causes of mitral stenosis?'
  - Rheumatic fever is the most common cause. Others include congenital lesions, carcinoid tumours and mucopolysaccharidoses.

2 'What are the complications of mitral stenosis?'
  - Pulmonary hypertension (leading to right-sided heart failure).
  - Haemoptysis.
  - Pulmonary oedema.
  - Emboli (stroke risk).
  - Atrial fibrillation.
  - Infective endocarditis.
  - Pressure effects from an enlarged left atrium: hoarseness (Ortner's syndrome, due to left recurrent laryngeal nerve compression), bronchial obstruction, dysphagia.

3 'What surgical procedures require antibiotic prophylaxis?'
  - Prophylaxis is generally not recommended now for surgical procedures.
  - Patients should be advised to maintain good oral health, and those at risk of infective endocarditis should be investigated and treated promptly when displaying evidence of infection.

Key points

- Mitral stenosis is a mid-diastolic murmur which is heard loudest at the apex in the left lateral position.
- Comment on the presence or absence of atrial fibrillation.
- Always check for the presence of pulmonary hypertension.

Mitral regurgitation

'Please examine this patient who has shortness of breath and/or palpitations.'

Findings

- General
  - tachypnoea, flushed face, peripheral oedema, mitral valvotomy scar
- Peripheral
  - raised JVP, irregularly irregular pulse
- Chest
  - inspection/palpation: scars (valvotomy, left thoracotomy), thrusting and laterally displaced apex, palpable thrill, parasternal heave
- Auscultation
  - pansystolic murmur loudest at apex radiating to the axilla, soft S1, S3 (+/–gallop) in severe MR

Presentation

'This patient has mitral regurgitation. The pulse is irregularly irregular consistent with atrial fibrillation. The apex beat is thrusting in nature. On auscultation there is an audible third heart sound and a pansystolic murmur, loudest at the apex, that radiates to the axilla.'

- Mitral regurgitation is often associated with ischaemic heart disease and a dilated left ventricle, so it is important to look for scars from previous CABG surgery, as well as any risk factors (tar staining on the fingers, signs of diabetes, hypertension, corneal arcus and xanthelasma).

Investigations

- ECG
  - LVH, left strain pattern, evidence of ischaemia, AF
- Echocardiogram ± TOE
  - assess LV systolic function and dimensions
  - Doppler flow studies: size and site of regurgitant jet

- additional features: vegetations, papillary muscle/chordae tendineae rupture, other significant valvular lesions
- CXR
  - normal, cardiomegaly, large left atrium, pulmonary oedema

Management

- Medical
  - management of congestive cardiac failure
  - rate-control therapy and anticoagulation for atrial fibrillation
- Surgical
  - when symptomatic, especially if evidence of cardiomegaly and raised end-systolic LV volume and resultant LV systolic dysfunction and dilatation
  - options:
    - repair/reconstruction (valvuloplasty) is preferable: carries lower operative mortality and no need for long-term anticoagulation
    - valve replacement: elective or for emergency treatment of acute severe incompetence (IE/acute chordae tendineae/papillary muscle rupture)
    - percutaneous options: valvuloplasty ring, MitraClip®

Questions

1 'What are the causes of mitral regurgitation?'

- Acute
  - chorda tendineae rupture: due to degenerative valve disease, trauma, infective endocarditis, rheumatic MV disease, MVP
  - papillary muscle rupture post myocardial infarction
  - papillary muscle dysfunction due to ischaemia
  - infective endocarditis
- Chronic
  - rheumatic heart disease
  - functional MR secondary to left ventricular dilatation: results in lateral displacement of the papillary muscles and retraction of the valve leaflets; dilatation or calcification of the mitral valve annulus may also result in MR
  - mitral valve prolapse
  - connective tissue disorders: SLE, rheumatoid arthritis, ankylosing spondylitis

- inherited disorders: Marfan's syndrome, Ehlers–Danlos syndrome, osteogenesis imperfecta, pseudoxanthoma elasticum
- hypertrophic cardiomyopathy

Key points

- Mitral regurgitation is a pansystolic murmur which is heard loudest at the apex and radiates to the axilla.
- Check for the presence/absence of atrial fibrillation.
- Be aware of the acute and chronic causes of mitral regurgitation.

Mitral valve prolapse

Findings

- Auscultation
  - normal heart sounds
  - mid-systolic click (may or may not be present)
  - late-systolic murmur loudest at the left sternal edge
- The murmur is made louder by factors that decrease the volume of blood within the cardiac chambers, i.e. straining (Valsalva manoeuvre) or standing from squatting
- These patients are often tall and thin with a higher female preponderance
- Patients are usually asymptomatic
- The patient may have peripheral stigmata of a connective tissue disorder

Investigations

- CXR
  - normal, cardiomegaly, large left atrium, pulmonary oedema
- ECG
  - may be inferior T-wave inversion, large left atrium, AF
- Echocardiogram
  - diagnostic!
  - degree of thickening of the mitral valve leaflets and their displacement relative to the annulus is indicative of MVP

Management

- Medical
  - palpitations controlled with beta-blocker therapy
- Surgical
  - indications as for MR; valve repair preferred to replacement

Questions

1 'What are the associations of mitral valve prolapse?'
  - Congenital heart disease: ASD, PDA

  - Congenital disorders: Turner's syndrome, Marfan's syndrome, osteogenesis imperfecta, pseudoxanthoma elasticum
  - Others: SLE
2. 'What are the presenting features of MVP?'
  - Usually asymptomatic but can be associated with atypical chest pain, palpitations, fatigue and dyspnoea
3. 'What are the complications of MVP?'
  - Infective endocarditis
  - Atrial and ventricular arrhythmias (ventricular ectopic beats are the usual cause of palpitations experienced)
  - MR
  - Cerebral emboli resulting in TIAs and/or stroke
  - Sudden cardiac death

Key points

- Mitral valve prolapse is a late-systolic murmur which is heard loudest at the left sternal edge.
- It is more common in females.
- Be aware of the associations of this murmur.

Mixed aortic valve disease

'This patient has a murmur. Please examine their cardiovascular system.'

Findings

- Features of both aortic stenosis and aortic regurgitation; the clinical findings should determine the predominant abnormality

Presentation

'This patient has mixed aortic valve disease with a predominant stenotic lesion. There is a slow-rising radial pulse and a normal blood pressure, but there are no stigmata of infective endocarditis. The apex beat is in the 5th intercostal space, midclavicular line. The heart sounds are normal, but there is a harsh ejection systolic murmur radiating to the carotids and an end-diastolic murmur heard loudest at the left sternal edge in forced expiration.'

- When suspecting mixed aortic valve disease, it is crucial to look for and present the above features. Be sure to ask about the patient's blood pressure. Where possible, decide upon the predominant lesion.

Investigations

- Diagnosis
  - Echocardiogram to assess the aortic valve, including valve size and gradient (also to assess the other valves, and left ventricular size and function)
  - TOE may be required to fully delineate the valvular anatomy.
  - Cardiac MRI scan can assist with severity and myocardial viability
  - Cardiac catheterisation can be used to assess for coronary artery disease; left ventricular angiography and an aortogram may further aid diagnosis
- Complications
  - ECG, echocardiogram, blood cultures if endocarditis suspected.
  - Exercise testing to assess exercise tolerance may be used to assess haemodynamic significance (care)

Management

- Medical management is an option with diuretics and possibly an ACE inhibitor (predominant AR) but surgery or TAVI should be considered. The indications for surgery are those for either specific valve lesion.

Key points

- When the murmur of aortic stenosis is detected, it is important to also listen for the diastolic murmur of aortic regurgitation (or any other murmur present).
- State which lesion is dominant in conjunction with the clinical findings.

Mixed mitral valve disease

'This patient has a murmur. Please examine their cardiovascular system.'

Findings

- Features of both mitral stenosis and mitral regurgitation; the clinical findings should determine the predominant abnormality

Presentation

'This patient has mixed mitral valve disease with a predominant regurgitant lesion. There are no stigmata of infective endocarditis. There is a sharp radial pulse. The apex beat is in the 5th intercostal space, towards the anterior axillary line. The heart sounds are normal with a pansystolic murmur radiating to the axilla and a mid-diastolic rumbling murmur heard loudest at the apex in the left lateral position.'

- When suspecting mixed mitral valve disease, it is crucial to look for and present the above features. As with mixed aortic valve disease, decide upon the predominant lesion.

Investigations

- Diagnosis
  - Echocardiogram to assess the mitral valve, including valve size and gradient (also to assess the other valves, and left ventricular size and function)
  - TOE may be required to fully delineate the valvular anatomy
  - Cardiac catheterisation further assesses valvular anatomy and concomitant coronary artery disease
- Complications
  - ECG (AF), CXR, echocardiogram, blood cultures if endocarditis suspected

Management

- Anticoagulation, rate-control therapy for atrial fibrillation and diuretics to treat fluid overload.
- Surgical valve replacement should be considered (as per the guidelines for individual valve lesions earlier in this chapter).

Key points

- Again, it is important to state which lesion is dominant.
- Look for features of endocarditis.

Prosthetic heart valves

'This patient attends for routine follow-up. Please examine their cardiovascular system.'

Findings

- General
  - dyspnoea, audible 'click' from end of bed
- Peripheral
  - scars (midline sternotomy, lateral thoracotomy) – including harvesting scars on legs/arms for concomitant CABG (more common in aortic valve replacement)
  - jaundice, anaemia, purpura
- Chest
  - aortic valve replacement: metallic S2 (opening/closing click), ejection systolic flow murmur (common, non-pathological), diastolic regurgitant murmur (pathological)
  - mitral valve replacement: metallic S1 (opening/closing click), systolic regurgitant murmur (pathological)
  - Note: tissue-valve replacements may demonstrate normal heart sounds (be alert to this possibility if sternotomy and no venous graft harvest scars evident!)

Presentation

'This patient has an aortic metallic valve replacement as evidenced by a midline sternotomy scar and a metallic second heart sound. The valve appears to be functioning well as I cannot hear a murmur of aortic regurgitation and there is no evidence of congestive cardiac failure. Of note there are no peripheral stigmata of subacute bacterial endocarditis.'

- Mention the type of valve replacement and whether there is regurgitation.
- Mention stigmata of endocarditis, any evidence of leg venous graft harvest scars for CABG and signs of over-anticoagulation.
- Look for any signs of congestive cardiac failure.

Investigations

- Echocardiogram

Management

- Metallic valves require lifelong warfarinisation
- Tissue valves require less anticoagulation but have a shorter lifespan

Questions

1 'What are complications of prosthetic valves?'
  - Early
    - complications of surgery
    - endocarditis
  - Late
    - thromboembolic sequelae
    - bleeding (as a result of anticoagulation)
    - infective endocarditis
    - valvular or paravalvular leak (regurgitation)
    - haemolysis (with anaemia)
    - valvular stenosis (endothelialisation (pannus))
    - structural failure/embolisation (very rare)
  - All of the above complications can occur with metallic valves. The main risk with tissue valves is regurgitation with degeneration (the risk of endocarditis in tissue valves is reduced).

2 'What influences the type of valve a patient will receive?'
  - Metallic valves are used in younger patients due to their durability (they are also used if the patient is already on warfarin), unless female and planning to conceive in the future.
  - Tissue valves are useful in the elderly (where the patient's life expectancy is thought to be less than that of the valve), and those with a risk of bleeding if warfarinised.

3 'What types of metallic valve are available?'
  - Ball valve (e.g. Starr–Edwards')
  - Disc valve (e.g. Björk–Shiley's)
  - Bi-leaflet valve (e.g. St Jude)
  - Ball valve devices cause a greater degree of haemolysis, whereas disc valves are more thrombogenic

7 'What happens to the intensity of the clicks from a metallic valve when failing?'
  - Decreasing intensity of the closing click

Key points

- Listen carefully from the end of the bed for any audible 'clicks'.
- Examine for any surgical scars.
- Comment on the type of valve (tissue/metallic), any evidence of infective endocarditis, and any evidence of congestive cardiac failure.
- Be aware of early and late complications of prosthetic valves.

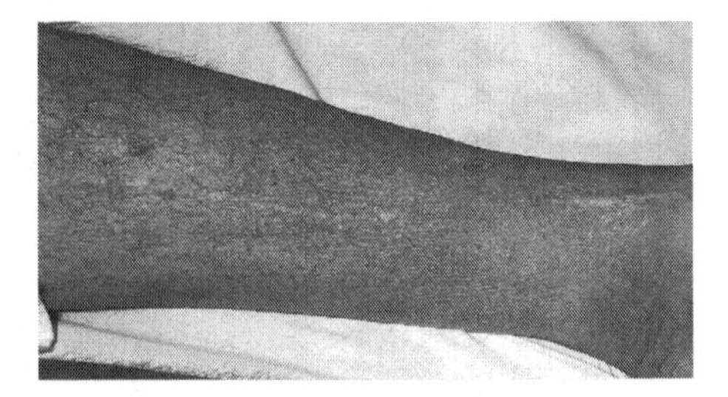

Figure 3.2 Vein harvesting scar on lower leg

Tricuspid regurgitation

'This patient has presented with shortness of breath. Please examine her cardiovascular system.'

Findings

- General
  - dyspnoea, jaundice
- Peripheral
  - marked peripheral/sacral oedema, prominent JVP (giant 'V' waves)
  - evidence of IVDU
- Chest
  - right ventricular heave, palpable P2, pansystolic murmur heard loudest at the left sternal edge in inspiration, S3, bi-basal crackles, evidence of chronic pulmonary pathology/ kyphoscoliosis
- Abdomen
  - ascites, pulsatile hepatomegaly

Presentation

'This patient has a diagnosis of tricuspid regurgitation as evidenced by the pansystolic murmur which is best heard at the LSE in inspiration. She also has a prominent JVP with giant 'V' waves. There is also evidence of pulsatile hepatomegaly which supports this diagnosis.'

- The main differential diagnosis for this murmur is mitral regurgitation, which would be louder in expiration and radiates to the axilla.
- Underlying causes can be divided into:
  - Primary
    - rheumatic heart disease, right-sided endocarditis (IVDU), carcinoid syndrome, penetrating trauma
  - Secondary
    - right-sided heart failure (e.g. pulmonary emboli, MI), pulmonary hypertension (cor pulmonale), Eisenmenger's syndrome.

Table 3.1 Auscultation findings of common valvular abnormalities

Lesion	Site	Timing	Character	Bell or diaphragm	Position	Respiratory phase
Aortic stenosis	Aortic area	ESM	Rough/harsh	Diaphragm	Forward	Expiratory
Aortic regurgitation	LSE → apex	EDM	High pitch/ decrescendo	Diaphragm	Forward	Expiratory
Mitral stenosis	Apex → axilla	Mid-diastolic murmur	Low pitch, rumbling	Bell	Left lateral	Expiratory
Mitral regurgitation	Apex → axilla	Pansystolic murmur	Blowing	Diaphragm	Left lateral	Expiratory
Tricuspid regurgitation	LSE	Pansystolic murmur	Blowing	Diaphragm	Forward	Inspiratory
Ventricular septal defect	LSE	ESM	Harsh/high pitch	Diaphragm	Forward	Expiratory

Investigations

- CXR
  - pulmonary congestion or pruned vessels, hyperinflated lungs, pulmonary fibrosis, cardiomegaly (large RA)
- Echo
  - demonstrates valvular abnormality, ventricular dimension and function and measures pulmonary artery pressures

Management

- Identify and treat the underlying cause
- Medical
  - diuretics, ACE inhibitors, spironolactone
- Surgical
  - valvuloplasty ring, valve replacement (rarely performed)

Questions

1 'What is carcinoid syndrome?'
  - Carcinoid tumours are neuroendocrine tumours of the enterochromaffin cells; they are mostly asymptomatic.
  - Carcinoid syndrome is evident when products of the tumour are metabolised by the liver.
  - Symptoms (bronchoconstriction, flushing, diarrhoea and CCF) are caused by release of serotonin, tachykinins and other vasoactive peptides into the circulatory system.

2 'How is carcinoid syndrome treated?'
  - Medical treatment
    - loperamide: symptomatic relief of diarrhoea
    - octreotide: somatostatin analogue; blocks the release of tumour mediators
  - Surgical treatment
    - curative resection
    - debulking of tumour

3 'What are the common organisms in infective endocarditis?'
  - *Streptococcus viridans*
  - *Staphylococcus aureus* (common in IVDUs)
  - *Staphylococcus epidermidis*
  - Enterococcus
  - Staphylococcal endocarditis has the worst prognosis, streptococcal the best

Key points

- Tricuspid regurgitation is a pansystolic murmur which is heard best at the left sternal edge in inspiration.
- Ensure to examine for peripheral signs such as pulsatile hepatomegaly and a prominent JVP.
- Always assess for signs of right heart failure.

Ventricular septal defect

'Please examine the cardiovascular system of this patient who has presented with dyspnoea.'

Findings

- General
  - well looking, often young
- Peripheral
  - possible clubbing (can occur in non-cyanotic congenital heart disease)
- Chest
  - inspection/palpation: laterally displaced apex, palpable thrill at LSE, left parasternal heave
  - auscultation: pansystolic murmur loudest at the left parasternal edge (also heard at the apex), loud P2 if pulmonary hypertension

Presentation

'This patient has a ventricular septal defect. The pulse is regular and there is no evidence of infective endocarditis. The apex beat is undisplaced. There is a left parasternal heave. On auscultation there is a pansystolic murmur loudest at the left sternal edge but also audible at the apex.' (comment on presence/absence of a loud P2)

- If the patient is young, the cause is likely to be congenital. In elderly patients look for factors contributing to ischaemic heart disease.
- Always look for evidence of complications or previous surgical repair.

Investigations

- ECG
  - likely to be normal if small asymptomatic VSD
  - other features include left-axis deviation and left ventricular hypertrophy in a moderate defect, or mainly right ventricular hypertrophy in a large defect
- Echocardiogram
  - assess size and position of VSD and degree of shunt

- CXR
  - pulmonary plethora (degree depends on size of VSD), cardiomegaly
- Cardiac catheterisation
  - shunt assessment through oxygen saturation measurement in the right ventricle

Management

- Conservative
  - most close spontaneously (especially if muscular defects)
- Surgical
  - surgical closure if: symptomatic VSD, infective endocarditis, post MI (acute septal rupture), haemodynamic compromise/volume overload, large shunt
  - may be closed percutaneously

Questions

1. 'What are the causes of a VSD?'
   - Congenital: 'lone', or associated with other defects, e.g. tetralogy of Fallot
   - Acquired: post MI (acute septal rupture), following trauma
2. 'What are the complications of a VSD?'
   - Infective endocarditis
   - Pulmonary hypertension
   - Eisenmenger's syndrome
   - Aortic regurgitation
3. 'What are the associations of a VSD?'
   - Aortic regurgitation
   - Patent ductus arteriosus
   - Coarctation of the aorta
   - Tetralogy of Fallot (VSD, overriding aorta, pulmonary stenosis, right ventricular hypertrophy)
   - Turner's syndrome
   - Trisomy 21 (endocardial cushion defects)

Key points

- The murmur associated with a VSD is a pansystolic murmur, heard loudest at the left sternal edge.
- Always check for a parasternal heave.
- Patients will often be young and asymptomatic.

Cardiology station summary

- The cardiac examination takes time; don't spend too long covering the peripheral signs, though at the same time be sure you don't miss any (a skill that can only be learnt with practice).
- Ensure that when you listen to heart sounds and murmurs, you are seen to time them with the carotids.
- If possible, when you present your case try to start with the diagnosis and explain why you think this is the diagnosis, and which investigations and management you would institute. This makes you sound confident, and if you really know your stuff you can demonstrate to the examiners that you are an excellent candidate without them having to ask any difficult questions.
- In valvular heart disease pay attention to signs other than the murmur (e.g. character of the pulse and apex beat and blood pressure); it is possible you may know the diagnosis before you listen to the chest.
- Be well rehearsed with the manoeuvres used to auscultate specific murmurs; this is an area where candidates often look clumsy.
- At the end of your examination remember to inform the examiners of the other pertinent examinations that you would like to perform, e.g. fundoscopy for Roth's spots in a case of subacute bacterial endocarditis, urinalysis (looking for evidence of haematuria).
- If you see a midline sternotomy scar ensure you look at the legs for evidence of venous harvesting scars. To not do this would be foolish; you can do it from the end of the bed before you have even touched the patient; you can tell that they have ischaemic heart disease and have had a coronary artery bypass graft.

Recommended reading

- Vahanian A, Baumgartner H, Bax J, *et al.* Guidelines on the management of valvular heart disease. The Task Force on the Management of Valvular Heart Disease of the European Society of Cardiology. *Eur Heart J.* 2007; **28**(2): 230–68.
- National Institute for Health and Clinical Excellence. *Prophylaxis Against Infective Endocarditis: NICE guideline 64*. London: NIHCE; 2008. http://guidance.nice.org.uk/CG64/Guidance/pdf/English (accessed 11 December 2010).
- National Institute for Health and Clinical Excellence. *Percutaneous Mitral Valve Annuloplasty: guidance. NICE guideline IPG352*. London: NIHCE; 2010. http://guidance.nice.org.uk/IPG352/Guidance/pdf/English (accessed 11 December 2010).

Station 4: Ethics and communication skills

Contents

Ethics station hints

- In this station your communication skills are being thoroughly evaluated.
- It is not just your ability to speak that is being assessed, but also your ability to listen to the concerns and issues that your patient presents.
- Read the scenario you are given, carefully. Look at what is being asked. The range of tasks in this station is wide, from breaking bad news to explaining a new diagnosis and addressing the related issues.
- Make good use of the time before entering the station to think through how you are going to tackle the scenario. Make a list of the issues to be addressed so that you do not forget them under the time pressure.
- Start with an open question. Give the patient as much time as they need to talk and allow them to finish before speaking.
- If this is an angry relative/patient that you are dealing with, always take on board their frustrations and apologise for any wrong-doing. Never disregard their concerns as the matter is important to them.
- Always try to empathise with the patient's situation.
- As the consultation progresses use more closed questions to help guide the consultation.
- Summarise the main points discussed and addressed, as you are going along, to act as a prompt to both you and the other individual for any other issues yet to be discussed.
- Remember that your ability to give accurate factual information, and the way in which you relay this, are both being assessed.
- Be clear and concise with your answers to any questions. Do not use jargon.
- During your discussion remember the four main ethical principles and use them to help guide your approach to any ethical dilemma:
  - autonomy: respect for the individual and the choices they make
  - justice: equality in the distribution of healthcare resources
  - beneficence: to act in the best interest of the patient
  - non-maleficence: actions should not harm the patient (based on the principle of *primum non nocere*: first do no harm).

- Remember the issue of confidentiality in this station. Confidentiality may be breached in situations such as: child protection issues, notifiable diseases, fitness to drive and serious crimes.

Breaking bad news

Background information

You are asked to see a 64-year-old lady in the respiratory outpatient clinic with the results of her staging CT. The patient was referred with a history of a persistent cough unresponsive to several courses of antibiotics. A chest X-ray requested by the GP prior to the referral showed a mass in the right upper lobe. The CT shows a mass in the right upper lobe, mediastinal and subcarinal lymph nodes, a small pleural effusion and multiple liver metastases. The current radiological staging is T3 N2 M1b, making this cancer inoperable.

Key points for the patient

- You have been seeing your GP over the past two months with a persistent cough.
- You have come to clinic to get to the bottom of this and don't want anyone else to be here with you.
- You have been given several courses of antibiotics which have not cleared the cough.
- You have been getting progressively more breathless over the past one month.
- Your children have noticed a change in your voice and you have lost a stone in weight recently.
- You have smoked 10 cigarettes a day for the past 40 years but stopped two years ago.
- You have had pain in your right arm and difficulty raising it.
- The GP recently started you on a blue inhaler and a course of steroids.
- You are annoyed that your GP has not taken you more seriously.
- You have a strong suspicion that you have cancer.
- The main concern you have is that your father died of lung cancer and was in a lot of pain towards the end.
- You want to be offered all possible treatment available and are infuriated when you are told that the treatment will be palliative and not curative.

- You want to lodge a formal complaint against your GP as things may have been different if you had been referred sooner.

Suggestions for the candidate

- Introduce yourself to the patient and ask the patient their understanding of why they have been referred to the clinic.
- Ask them if they have come with anyone today and if they would like anyone else present.
- Ask the patient about the symptoms they have been having, what prompted them to see their GP, and ask about any treatments they have had.
- Ask about their main concerns and what they think might be going on.
- Recap the history to the point where investigations began and tell the patient that you have all of the results.
- The main reason for investigation was to exclude a malignancy.
- Unfortunately the results show that there is a tumour in the lung involving lymph nodes and lesions in the liver; the diagnosis is likely to be cancer.
- Pause for the patient to take the information in and give them as much time as they need.
- Ask the patient if they have understood what you have told them, before proceeding.
- Explain that the case has been discussed at the MDT meeting.
- Explain that because of the spread, the cancer can't be surgically resected and that treatment options will be limited; however, you can offer the patient hope and say that they can be referred to oncology for a further opinion.
- Ask the patient what their main concerns are and acknowledge them.
- Reiterate that although it is not possible to treat the underlying cause there are lots of services and support that can enable them to cope with the diagnosis and the disease.
- Acknowledge the patient's anger and frustration with the GP but advise them that a diagnosis of cancer is sometimes difficult to make, and that they ought to contact the GP to discuss the issues.
- Tell the patient that they can see a Macmillan nurse today before they go, and that they will be a regular point of contact.
- Recap the information given and ask if the patient has any questions.

- Explain that you will see them in clinic next week and ask if they would like to bring anyone along.

Themes explored

Breaking bad news for an incurable condition

- This is often difficult due to the emotions involved and the range of patient responses that one may encounter. Remain calm yourself and make sure that you listen to the patient.
- Ensure that you ask the patient about their understanding of the current situation before proceeding to give them the news.
- Do not use any euphemisms or jargon and make sure that you are being clear with the diagnosis.
- Make sure that you have all the correct/relevant information to hand.
- Appear empathic and ensure that the patient has time to digest the information given.
- Remember to summarise the key points at the end of the consultation and offer to see them again in clinic.
- Offer the relevant support: Macmillan nurses, palliative care teams, information leaflets and contacts for support groups.

Dealing with an angry patient

- Acknowledge that the patient is angry or upset and let them know that you are aware and that it is a normal response.
- Allow them to voice their anger and do not react to the situation.
- Ask them what specific things are making them angry and offer them potential solutions to help overcome these issues.

Driving regulations

Mr David Clegg has come to see you in the sleep clinic. He is a 55-year-old gentleman who has been complaining of symptoms of tiredness over the past year. He has been investigated by his GP and recent thyroid function tests and a full blood count have been normal. The patient is complaining that his symptoms are interfering with his work and he is falling asleep inappropriately. The GP is concerned that the patient may have obstructive sleep apnoea/obesity hypoventilation syndrome. The patient weighs 144 kg and is 1.65 m tall. Discuss the diagnosis and related information with the patient.

Key points for the patient

- You have been referred to the clinic by your GP to investigate the cause of your sleepiness.
- All your blood tests have been normal.
- Your GP thinks that the tiredness may be related to the fact that you are rather overweight.
- The tiredness is having a significant impact on your life.
- You have fallen asleep in several important meetings at work, which was rather embarrassing.
- You are concerned as you have found yourself falling asleep at the wheel of your car. You nearly crashed into a barrier the other day.
- Your wife sleeps in a separate bedroom and this is affecting your sex life as well.
- You have tried losing weight but you are not getting very far.
- You drink up to three pints of beer most nights.
- You wake up most mornings complaining of a headache.
- You feel as though you have not had a good night's sleep for months.
- You are beginning to feel depressed about this now.
- You want to know what can be done.

Suggestions for the candidate

- Start by asking the patient if he knows why he has been referred to the sleep clinic.
- Ask him specifically how the tiredness is impacting on the various aspects of his life; ensure that you ask about work and home.

- Ask about a bed partner and if they have reported the patient snoring at night or having periods where they stop breathing.
- Ask the patient what they understand about the terms obstructive sleep apnoea and obesity, and whether the GP has explained anything about the diagnosis.
- Go on to explain the potential diagnosis.
- Explain to the patient that further tests will have to be carried out before confirming the diagnosis.
- Initially the patient will be asked to fill out an Epworth Sleep Score and then further tests will include overnight oximetry and/or full polysomnography.
- Advise the patient on lifestyle factors: cut down on alcohol intake and attempt to lose weight.
- One of the main issues that will need to be covered here is that the patient is falling asleep at the wheel of his car.
- You will have to advise him that he will need to stop driving and inform the DVLA. The regulation for group 1 drivers is that they need to stop driving until satisfactory control of symptoms has been attained.
- Ask the patient if they have any questions.
- Summarise the above information for the patient before ending the consultation and offer to give them some information leaflets prior to leaving.

Themes explored

- The main points to cover in this case are to give the patient information regarding the potential diagnosis and also about what investigations may be required.
- The patient has specifically mentioned that he is falling asleep whilst he is driving. This needs to be taken seriously and it is the responsibility of the medical practitioner to advise the patient to stop driving and inform the DVLA. This should also be documented in the notes in a real-life situation.
- There are many medical conditions about which the patient needs to inform the DVLA and may also need to stop driving.
- Initially the patient should inform the DVLA; however, if it comes to a point that after repeated advice the patient still has not done this, then one may need to break confidentiality and report this to the DVLA.

Table 4.1 Medical conditions and driving regulations

Condition	Group 1 regulations	Group 2 regulations
First seizure	6 months off driving from 1st seizure	5 years off driving from 1st seizure
ICD implantation (prophylactic)	1 month off driving from implant	Permanently barred
Insulin-treated diabetes	Must recognise symptoms of hypoglycaemia – 1, 2 or 3-year licenses given	Permanently barred
Percutaneous coronary intervention (elective)	1 week off driving	Disqualified for 6 weeks until exercise test requirements fulfilled
CVA/TIA	Stop for 1 month – no residual defect before restarting	License revoked for 1 year

Notes

Group 1: Car/motorcycle

Group 2: Heavy goods vehicle/public carriage vehicle

Initiating a new therapy

You are the medical registrar in a rheumatology outpatient clinic. Mrs Patel is a 52-year-old lady who has a history of severe rheumatoid arthritis. She has been tried on several disease-modifying drugs (including methotrexate) without significant reduction in disease activity. She has come to see you in clinic today to discuss the possibility of starting anti-TNF therapy.

Key points for the patient

- You have suffered with severe rheumatoid arthritis for many years now.
- You have active disease which is proving difficult to control and is affecting your quality of life and ability to work as a shop assistant.
- Steroids have caused significant thinning of your bones so you are reluctant to have the dosage increased.
- You have tried several DMARDs including methotrexate, which caused a problem with your liver tests so had to be withdrawn.
- You are fed up with repeated trials of medications which either do not work or cause more problems due to side effects, and have read about Remicade (infliximab) on the internet and insist that you are treated with this.
- You understand this is a new and expensive treatment but insist that this is the treatment you want.
- You want to know more about how these drugs work; will they cure you?
- You want to know what the potential effects of the drugs are; importantly, will they damage your liver like methotrexate.
- You are afraid of the immunosuppressive effects of the drug as you had a TB scare about 20 years ago, back in India, when a lump was found in one of your neck glands not long after your mother was diagnosed with TB.
- You do not recall whether or not you received any treatment at the time, only that you had a biopsy taken.
- You are concerned that the history of TB will affect your eligibility for treatment with infliximab and want to know what can be done to ensure that initiation of treatment is not delayed.

Suggestions for the candidate

- You have a very knowledgeable patient in clinic with you, who is very well read about her condition.
- You ask how she is managing at present with her condition, and how it is affecting her life.
- Listen to and empathise with the patient's concerns regarding the effects of the condition and the impact it has on her.
- You note that the patient has tried several DMARDs in the past without much benefit.
- You explain that abnormalities with liver function tests are a well-documented effect of treatment with methotrexate and unfortunately osteoporosis is one of the long-term effects of treatment with steroids.
- You enquire where the patient read about Remicade – an official evidence-based website or from a general internet search.
- You explain that infliximab is a new treatment and that there are strict criteria for eligibility; however, she appears to have met the criteria, having failed on more than two DMARDS, one of which was methotrexate.
- Explain that the medication works by reducing inflammation but cannot cure the disease.
- Give the patient time to take this in and ask any questions.
- Explain that these medications can in very rare cases cause serious liver damage; however, more commonly, as they dampen the immune system, they can be associated with serious infections such as TB. Enquire about past TB exposure.
- Before initiating therapy, TB needs to be excluded with a chest radiograph; if it is positive, and as there is uncertainty whether the TB was adequately treated previously, she will need chemoprophylaxis before treatment with infliximab.
- Reassure her this is usual practice; that it is safer to delay starting infliximab for a short time to prevent potential life-threatening infection with TB, and that this does not mean she will be precluded from receiving treatment.
- Make sure you listen to and address the patient's concerns; summarise and ask if there are any questions.
- Offer an information leaflet about infliximab and schedule another appointment to discuss any concerns and initiate the treatment pathway.

Themes explored

- Autonomy: the main theme explored in this case is how to counsel a patient effectively regarding a new treatment, without coercion, to enable them to make an informed decision as to whether or not to proceed with the treatment.
- Beneficence: with novel therapies, there is a risk–benefit balance between the potential adverse effects of the drug versus the desired benefits, which is often difficult when the long-term effects of such medications are often not fully known.
- Justice: NICE guidelines suggest that patients with active RA who have failed treatment with at least two DMARDs, one of which should have been methotrexate (unless intolerant), should be considered for anti-TNF therapy. This would justify the need to give such potentially harmful treatments.
- Do No Harm: prior to treatment with anti-TNF therapy, patients should be screened for TB, and active TB must be adequately treated, as therapy carries an increased susceptibility to developing TB.
- Patients with a past history of extra-pulmonary TB or abnormal CXRs require close monitoring on treatment.
- Patients with previously inadequately treated TB require chemoprophylaxis before commencing treatment.
- Patients require close monitoring for symptoms of TB whilst receiving anti-TNF therapy and for six months after stopping.
- If patients develop symptoms suggestive of TB on anti-TNF therapy they will require full treatment with chemotherapy.

Long-term condition

You are the registrar in the renal outpatient clinic. You are seeing a 32-year-old female patient who has recently been given a diagnosis of adult polycystic kidney disease. She has come back to clinic today wanting more information regarding her diagnosis and the implications it may have for her future. Her mother also has the condition and may be starting dialysis in the near future. She has a young daughter and is planning to expand her family.

Key points for the patient

- You were recently seen in the renal clinic when you were given a diagnosis of polycystic kidney disease.
- You have had some time to digest the information given to you previously and you now have several concerns that you would like to discuss with the doctor today.
- You are extremely anxious regarding the diagnosis.
- You have a three-year-old daughter and had been planning to expand your family.
- You are concerned regarding the prognosis of the condition and associated complications.
- You enquire if there are any alternative treatments or surgical options available to help prevent any complications and decline in renal function.
- Your mother also has the disease and her doctors are considering starting her on dialysis.
- You wish to know whether this would be the case for you in the future and if so when might this be, and whether would you require a kidney transplant at some stage.
- You wish to know if there is anything that can be done to prevent yourself from getting to the stage your mother is currently at, i.e. requiring long-term renal replacement therapy.
- As your family is not yet complete, you wish to know if this diagnosis would prevent you from having any more children and what are the risks involved.
- If you were to become pregnant, would pregnancy affect the progress of the disease?

- Since both you and your mother have the condition, you are extremely concerned about the inheritance of the condition and whether you could have passed on the disease to your daughter.
- You want to know if your daughter can be tested and if there is anything that can be done to prevent her from getting the condition.

Suggestions for the candidate

- It is important firstly to establish the amount of information the patient received during the previous consultation and how much she has retained.
- Summarise the key points and offer to discuss the diagnosis again, and clarify any points that she may not have fully understood.
- Ascertain the patient's main concerns by asking if she has any specific questions in mind that she would like answered.
- Attempt to alleviate some of the patient's anxieties by reassuring her that you and the team are there to help and provide whatever support she requires.
- Explain again to the patient that APKD is an inherited condition (autosomal dominant) and that means that there is a 50% chance that she has passed it on to her daughter.
- Explain that as it is a cystic condition, the cysts could develop in other organs (liver, pancreas) as well as the kidney. Problems she might experience include recurrent urinary tract infection, infection or bleeding within a cyst, and high blood pressure, all of which are treatable.
- Explain that end-stage renal failure is a potential complication that may occur some years after diagnosis. If her renal function were to decline over the course of time then she would be prepared for dialysis in advance of needing it. An alternative possibility would be transplantation when a kidney became available; however, she may require dialysis initially.
- Explain that the mainstay of treatment is trying to preserve kidney function for as long as possible by reducing complications and treating them aggressively, i.e. good blood pressure control, treating infections, etc.
- Explain to the patient that she will be closely monitored over the coming years with regard to her renal function.
- Reassure the patient that this disease will not prevent her from having more children, but that in future pregnancies, she will be monitored more closely and will have an obstetrician-led pregnancy with close monitoring of her renal function.

- Explain that this will mean more frequent prenatal visits to the hospital and scans to ensure that she and her baby are both doing well. In addition she will be more closely monitored during and after delivery.
- Tell her that pregnancy per se will not affect disease progression.
- Explain that the main way of testing for this disease is using ultrasound imaging to look for the presence of cysts within the kidney. This is usually done when the patient is in their 20s when the disease itself manifests.
- When asked about genetic testing, explain that there are many genetic mutations that can give rise to the disease, so it may not be possible to identify the specific one causing her disease. Assure her that her daughter will be under surveillance as she gets older.
- Summarise that although PKD is a chronic/lifelong condition it will not prevent her from leading a full and active life. She will be closely monitored throughout with specialist services when needed. It will not prevent her from completing her family and should she progress to needing dialysis she will be thoroughly supported and helped through this.

Themes explored

- When discussing a new diagnosis of a chronic/life-limiting condition it is important to ascertain the level of information the patient has received beforehand.
- Respecting the patient's autonomy is crucial and it is important to gauge the level of information the patient would like to receive.
- Offer the patient details of support groups where they can obtain further advice in dealing with the diagnosis, and provide information leaflets that explain the condition.
- Explain that having a chronic disease does not mean that she cannot complete her family.
- She must be informed that 50% of her offspring may also be affected with the condition due to the pattern of inheritance.
- The importance of reducing complications must be emphasised. If renal decline is progressive then development of end-stage renal failure may be unavoidable, though there are options of renal replacement therapy.
- Offer a further consultation to discuss any points and referral to an obstetrician, if desired, for future pregnancy planning.

Medical error

You are the registrar covering the medical wards for the weekend. You have received a call from one of the nurses on the cardiology ward to speak to a patient's daughter. The patient's daughter is an ITU nurse and is very concerned about a large haematoma on the left side of her mother's neck following the attempted insertion of a central line. The patient required a central line for the administration of amiodarone.

Key points for the daughter

- You have come to the hospital to see your mother, who was admitted with an NSTEMI.
- You are very distressed to see that she has developed a large haematoma on the left side of her neck, although it does not appear to be bothering her.
- On speaking to the nursing staff, you learn that your mother had a central venous catheter inserted last night so that amiodarone could be administered to control new-onset fast atrial fibrillation.
- You ask the nursing staff to look for any documentation as to whether the insertion may have been traumatic or difficult.
- When the medical notes are checked, the cardiology nurse informs you that the first attempt at insertion was unsuccessful. You enquire as to why this was and discover that ultrasound guidance was not used.
- You are very angry at the medical staff and demand to know how your mother has come to develop a haematoma.
- You ask the doctor why a peripheral cannula could not have been used to administer the drug.
- You ask the doctor why the central line was not inserted under ultrasound guidance, as after all, this is the national guideline.
- You think that it is completely unacceptable that recommended guidelines have not been adhered to and wish to know the name and grade of the doctor who did the procedure.
- You want to speak with the consultant in charge of your mother's care and want to make an official complaint about the incident as you are concerned that it was bad practice.

- You also insist that an incident form is completed regarding the event to ensure the matter and the individuals involved are dealt with appropriately.

Suggestions for the candidate

- Note that you are dealing with an angry relative who is a fellow health professional.
- Allow the relative to voice all their concerns and provide adequate time for them to do so.
- Do not interrupt.
- Acknowledge the daughter's concerns.
- Explain that her mother developed acute-onset fast atrial fibrillation, which was difficult to control and therefore required the use of amiodarone. The amiodarone needed to be given via central access.
- Explain that the haematoma occurred as a result of an attempt at central line insertion.
- Be open and honest and acknowledge the fact that as the line was not inserted using ultrasound guidance, this may have accounted for the failure of insertion and the resultant haematoma.
- Apologise for the fact that it was not done under ultrasound guidance and for any harm that the patient has sustained.
- Assure the daughter that the fellow registrar who inserted the central line was certified as being competent in performing the procedure.
- Address the daughter's concerns regarding the ultrasound and explain that when inserting central lines it is normal practice to insert them under ultrasound guidance as per NICE guidelines.
- Explain that the possible reason for the procedure having been performed without the aid of ultrasound was that it was not available, and it was important to get the line in as quickly as possible to administer the drug.
- A lengthy delay in inserting the line could have resulted in her mother becoming more unwell with her fast irregular heart rate.
- Explain that it is departmental policy to give amiodarone through a central line due to the potential side effects of peripheral administration.
- Offer to arrange a time for the daughter to speak with her mother's consultant.

- Reassure her that you will fill out an incident report regarding the matter.
- Advise the daughter that if she or her mother would like to take the matter any further, they can contact the patient liaison office and write a formal complaint, which would be addressed in accordance with hospital policies.
- Once again apologise for any distress caused to the patient and assure the daughter that you will relay her concerns to the various parties involved.

Themes explored

- The two ethical principles addressed in this case are beneficence and non-maleficence.
- The central line was inserted to aid the patient's treatment and prevent potential side effects of the drug from peripheral administration. However, by not using an ultrasound probe, this may have resulted in the patient being unduly harmed.
- When addressing any form of complaint, it is important that you have on hand as much information as possible to give to the patient or relatives.
- You should acknowledge any error that has been made and you must apologise for this.
- Offer to answer any questions that the patient or relative may have in relation to their concerns, to help alleviate the situation.
- Do not try to conceal any information which may later come to light.
- If they are not happy with your explanation, offer to arrange for them to speak with the consultant in charge of the patient's care.
- Offer them other sources of help regarding patient welfare such as patient advice services, which are available in all hospitals.
- Reassure them that the matter will be taken seriously and dealt with appropriately.

Mental capacity

You are the medical registrar on call over the weekend. You have been asked to speak to a relative on one of the medical wards, who wants her husband discharged against medical advice. Thc patient has a background history of Parkinson's disease, epilepsy and vascular dementia. The patent was admitted on this occasion with a lower respiratory tract infection and a UTI. He is currently having intravenous fluids and antibiotics and has systemic signs of sepsis. The patient's wife is adamant that she is going to discharge him from hospital. You try to speak to the patient but he is too confused to communicate.

Key points for the patient's wife

- You are the patient's main carer and have looked after him for many years now.
- You are adamant that he should be discharged, albeit against medical advice.
- You feel that his condition is worsening as he is not in a familiar environment.
- You also feel that there is no point him being in hospital as he is not getting the one-to-one care which he would be getting at home.
- You question the doctor as to whether he can allow the patient to have antibiotics at home.
- You know that the patient has signed an advance directive stating that he should not get treatment that would unnecessarily prolong his life.
- You feel that the patient is not getting sleep at night as it is too noisy.
- You know that you do not hold lasting power of attorney status, but you know the patient best and this is what he would want.
- You are angry at the doctor because you think nothing is being done as it is the weekend.
- You question the doctor as to what he can do to stop you from taking your husband off the ward.
- You feel that keeping him in hospital is not in his best interest.
- You can give him antibiotics orally and make sure he drinks lots of water at home.

- You are frustrated, as you feel that your views are not being taken into consideration.

Suggestions for the candidate

- You are approached by a very angry and frustrated relative.
- Make sure you listen and address the relative's concerns.
- Ask her what is worrying her regarding the patient's stay in hospital.
- Reiterate that the patient is suffering from a urinary tract infection and a chest infection, and is currently having intravenous antibiotics and fluids.
- Explain that you are part of a team and that you are all acting in the patient's best interest.
- When the relative tells you about the advance directive you can tell her that this infection is a treatable condition and you are not looking at end-of-life care yet.
- Explain to the relative that you will try to ensure that everything possible will be done to make the patient more comfortable, such as moving him to a side room or a quieter area of the ward where he should be able to get more rest.
- When the patient's relative asks what you can do to stop her, tell her that you would hope to resolve the situation before this happened, but if it did you would have to involve hospital security.
- Explain that you have tried to assess the patient's mental capacity but he is too confused to understand what you are saying.
- Ask the relative if she has lasting power of attorney.
- Explain that as she does not have this status, as the healthcare professional you are acting in the patient's best interest by keeping them in hospital.
- You would be going against your duty of care by allowing the patient to leave.
- Summarise what you have discussed and ensure that you convey an empathic tone to the relative and acknowledge her concerns.
- Suggest that the relative should speak to the consultant in charge of the patient's care, after the weekend.
- Reiterate that you are doing your best to treat the patient.

Themes explored

- The main theme explored in this case is the issue of mental capacity. The Mental Capacity Act came into force in 2007 in England and Wales and it helps provide a framework to make decisions for people who are unable to do so for themselves.
- To assess whether a patient has capacity, they must be able to understand and retain the information presented to them, weigh up the information, and communicate the decision back to you. If a patient does not have capacity then as the healthcare worker responsible for the patient you can take decisions regarding the patient's medical care. It is also important to involve the family and carers and seek their views and opinions as to what the patient may have wanted.
- Patients can sign advance directives refusing specific treatments that they would not want should the situation arise. These are legally binding documents that need to be signed in the presence of a witness. They should also contain a statement that the decision should be enforced even if life is at risk.
- If there is any issue regarding the validity of an advance directive in an emergency situation, the appropriate treatment can be given until the validity of the document is verified.
- The patient can also appoint a lasting power of attorney who can make medical (and other) decisions on behalf of the patient who lacks capacity. This document needs to be registered with the Office of the Public Guardian for it to be valid.
- In a situation where a patient lacks capacity and has no representative to discuss their medical treatment, an Independent Mental Capacity Advocate (IMCA) can be appointed to act as an advocate for the patient in the decision-making process. An IMCA is not needed if it is an emergency treatment or if the patient is detained and being treated for a mental illness under the Mental Health Act.

New diagnosis

You are the registrar in the gastroenterology outpatient clinic. Your next patient is a 19-year-old man who has recently presented with several episodes of bloody diarrhoea. The patient has had a recent colonoscopy which confirms the diagnosis of ulcerative colitis. Discuss with the patient the diagnosis and the implications of the disease.

Key points for the patient

- You have come to clinic today to find out the results of a colonoscopy that was undertaken to investigate the cause of the bloody diarrhoea you have been having.
- Previously, you have been fit and well with no other health problems. Over the past six weeks, you have had multiple episodes of bloody diarrhoea daily.
- When you were previously seen in clinic, the doctor mentioned that they were looking to see if there was any inflammation of your bowels that could account for symptoms. However, you did not really understand the explanation given.
- When you are given the diagnosis of ulcerative colitis, you ask the doctor to fully explain this and any problems that are associated with the disease.
- You are worried as you have just finished your gap year and are due to start university soon and wonder how this will affect your social life.
- You want to know what medication you will need to take.
- You want to know if you will still be able to drink alcohol.
- You want to know how often you will need to see a doctor.
- You read an information leaflet in the waiting room and you want to know if you will need to have an operation and more importantly if you will end up with a stoma.
- When pressed by the doctor, tell them that you are very concerned about this problem as your dad died at a young age from bowel cancer and wonder if this is something that you should be worried about, now that you have this diagnosis.

Suggestions for the candidate

- Recap the history for the patient and initially gauge their understanding of why they have come to clinic and why they have had the procedure done.
- Ascertain the amount of information previously given and their level of understanding.
- Summarise the previous information.
- Explain that colonoscopy is the usual investigation to look for any active inflammation of the bowel that could be causing his symptoms.
- Explain that biopsies were also taken at the same time to help clarify the underlying disease/inflammatory process.
- Explain that all the results are now available and are consistent with a diagnosis of ulcerative colitis.
- Give them time to absorb the information that they have been given and then wait to see if they have any questions.
- Ask if they have heard of this condition before and if so what is their understanding of it.
- Offer to give them an overview of the disease. Do not use jargon.
- Explain that UC causes inflammation of the bowel wall. It is restricted to the large bowel and therefore results in bloody diarrhoea, so his presentation was not unusual or atypical.
- Although the disease mainly affects the bowel there can be extra-gastrointestinal symptoms such as joint aches, rashes, mouth ulcers and sore eyes but bowel symptoms are commonest.
- Empathise with the patient and reassure him that with treatment there is no reason why he cannot lead a normal life and pursue his studies, and that the condition could be looked after by a specialist in his university town if needed.
- As he has only bowel symptoms, treatment would initially be local therapy with oral ASAs and/or steroids if symptoms persist.
- If symptoms are still not well controlled, stronger therapies to dampen the immune system could be added at that point.
- Explain that the amount of alcohol he can drink will depend upon the treatment he is taking; if on stronger immunosuppression it would not be recommended.
- Explain that initially he will need a more regular review (every couple of months or so) until the disease is under control but thereafter the intervals between clinic reviews would be increased.

- Reassure him that in between visits there is always a specialist nurse available who he can contact for further advice or earlier review if required.
- Regarding surgery, reassure him that this is undertaken only in severe cases of colitis when the inflammation cannot be controlled with medical therapy, or in emergency circumstances. However, there are now many new treatments available (biological) that can help avoid surgical intervention altogether, even in circumstances where it might have been unavoidable in the past.
- Ask the patient if they have any other specific concerns or if there is anything at all they are worried about. Tell them it is quite natural to be frightened and worried about the new diagnosis, and that you are here to help/address any concerns they may have.
- When bowel cancer is mentioned, be honest and tell him that in patients with UC, there is a slightly increased risk of bowel cancer after having the condition for many years. The risk however is still small. Emphasise that all patients are closely monitored and there are surveillance measures to pick up any disease early.
- Summarise the main points and emphasise that there is plenty of support available to help him cope with the condition, and the aim of treatment is to help him lead as normal a life as possible.

Themes explored

- When giving the diagnosis of a new/chronic/life-limiting condition the patient has the right to know the full facts regarding the diagnosis and all treatment options available.
- The patient's autonomy regarding their treatment decisions must be respected if they are of sound mind, especially any refusal of treatment.
- The individual must be allowed to use their own judgement in weighing up the pros and cons of available options. They should never be coerced into making any choices.
- It is crucial for the information giver to not allow any pre-formed judgements about the patient to cloud the information given to the patient.
- If there are any errors or delays in getting to the underlying diagnosis, these should not be hidden and any apologies should be made to the patient and their family.

- In this scenario it is a young patient who is clearly frightened as to what the future holds with this condition (regarding treatments including surgery, the possibility of cancer and whether he will still be able to lead a normal life). Reassure the patient constantly and give them plenty of sources of further information. Allow them time to digest the information and address their concerns as they arise.

Organ transplantation

You are a doctor working on the intensive care unit. You have been asked to discuss the possibility of organ donation with the family of an ICU patient who has been pronounced brainstem dead. The patient is a young, fit and healthy 19-year-old girl who was involved in a road traffic accident. The patient's father is her next of kin and you approach him to discuss the possibility of donation.

Key points for the father

- Your daughter has been involved in a road traffic accident, in which she was hit by a car.
- She suffered extensive head injuries and also had a ruptured spleen.
- You were informed by the consultant in charge of her care that her injuries were very severe and that she may not survive. You were naturally extremely upset as your daughter was young, fit and healthy.
- You were informed that if she deteriorated clinically, tests would be carried out to assess the condition of her brainstem.
- The nature of the tests was fully explained to you and subsequently, after two rounds of tests, your daughter was pronounced brainstem dead.
- As a family you are all in shock as the last time you saw your daughter before the accident she had been excited about her plans for her gap year.
- When asked by the doctor about the possibility of organ donation; you are initially taken aback that this has been raised at your time of loss.
- After some time to think, you ask what this would involve.
- You recall that your daughter had previously mentioned that she had ticked on her driving licence application that she would wish to be an organ donor.
- You want to know what the process of donation would involve and how long it would take.
- You ask which organs are likely to be taken.
- You ask who would receive the organs and whether you would be able to contact them.

- Your main concern is whether your daughter would experience any pain when the organs were removed.
- You are also worried about the appearance of her body after the organs have been taken.
- You ask if there is anyone else who can give you any further information about the process.
- You answer on the family's behalf that despite your tragic loss, you are willing to give permission for your daughter to be an organ donor, as this was her wish.

Suggestions for the candidate

- Initially offer your condolences to the family and empathise with their loss.
- When you bring up the subject of organ donation, acknowledge what a difficult time this must be for them, but explain that the possibility of organ donation is regularly addressed on ICU.
- Ask the family if they are aware of any wishes their daughter may have had concerning organ donation. Specifically ask about an organ donation card, or whether it was recorded on her driver's licence.
- Explain the process ahead. You will contact the organ transplantation team who will coordinate the process of organ removal. There will be various teams from around the country who will come to obtain the organs as this hospital is not a transplant centre.
- Advise that you will need to seek the permission of the coroner before organ retrieval can take place.
- The organs will go to individuals on a transplant waiting list, who have been identified as being suitable and most urgently in need of an organ transplant.
- As the patient was previously healthy, all the major organs could be used as well as other body parts such as corneas, tendons and possibly bones.
- Reassure the family that the patient will not feel any pain as she is brainstem dead and therefore unable to feel pain, but that she will also be given opiates prior to the procedure to ensure this.
- The surgeons will carefully close the body after removing the organs so the patient's appearance will not be significantly disfigured.

- The family will not be able to contact the recipients of the organs, but the transplant coordinator would be able to tell them where the organs have gone to.
- The transplant team/coordinator will also be available to answer any further questions that the family may have.
- Empathise with the family again regarding their loss and thank them all for considering this decision at such a difficult time.

Themes explored

- Organ transplantation is an area which is often very difficult to address as there is so much emotion involved when broaching the subject with relatives who have just lost a loved one.
- The legal aspects to remember are that the coroner's permission must be sought before going ahead with retrieval.
- If the next of kin disagrees with donation, despite the patient being on the organ donation register, it is current practice not to proceed with the donation regardless of patient autonomy.

Ethics station summary

- Thoroughly prepare for the range of themes commonly encountered in this station and this will stand you in good stead.
- Remember your communication skills are being challenged, and inadequate preparation will be evident to the examiners.
- Use the time before entering the station to plan a structure for your discussion and the topics to be addressed. This will help you organise your thoughts and think through the information to be gathered and given to the patient.
- Always bear the four main ethical principles in mind when approaching the task.
- Remember that there are other issues that may be the theme of discussion also, such as:
  - confidentiality
  - driving regulations/DVLA medical guidelines
  - breaking bad news
  - relaying a new diagnosis
  - organ transplantation
  - cardiac resuscitation orders
  - suitability for ICU transfer/care.
- Allow plenty of time for pauses to give the individual enough time to take the information on board and to ask any questions they might have.
- Summarise the key points of discussion as you go along.
- Remember the station carries many marks and may be the decider of your overall pass/fail, so practise, practise, practise!

Recommended reading

- Department for Constitutional Affairs. *Mental Capacity Act 2005*. London: HMSO; 2005. www.legislation.gov.uk/ukpga/2005/9/contents (accessed 12 December 2010).
- www.organdonation.nhs.uk/ukt/default.jsp (accessed 12 December 2010).
- National Institute for Health and Clinical Excellence. *Rheumatoid Arthritis: Adalimumab, Etanercept and Infliximab. NICE guideline TA130*. London: NIHCE; 2010 : http://guidance.nice.org.uk/TA130/Guidance/doc/English (accessed 12 December 2010).
- DVLA medical guidelines – www.dft.gov.uk/dvla/medical

Station 5:
Brief clinical encounters

Contents

Hints for station 5

- Station 5 is worth many marks in the exam and it is crucial to dedicate a substantial proportion of revision time to it.
- Carefully read the instructions to candidates before entering the station and brainstorm the specific questions you need to ask to help get the diagnosis. Formulate a structure for your focused clinical assessment.
- On entering the room, look closely for any clues to the diagnosis; 'spot diagnoses' are common in this station.
- Generally begin with an open question, but quickly become more focused after this.
- Establish a likely diagnosis (and differential) then find out about specific complications.
- Examination generally involves looking for specific signs to clarify a diagnosis rather than following a set pattern.
- Be sure to examine for evidence of complications/manifestations of the primary disorder.
- When presenting the case, give the likely diagnosis first, but have a differential available in case you are asked for this.
- Always ask about any specific concerns that the patient has, and provide a solution or explanation as appropriate. Addressing the patient's welfare and concerns is vital in this station.

Acromegaly

'This patient has been told he has been snoring heavily at night and has recently developed polydipsia. Please ask any relevant questions and proceed as appropriate.'

Focused history

- Duration of symptoms
- Evidence to suggest OSA: snoring, daytime somnolence, headaches, decreased libido
- Change in appearance: prognathism, macroglossia, increased hand span/shoe size (rings/shoes no longer fitting), large nose
- Metabolic symptoms; glycosuria (polydipsia, polyuria), increased sweating, tiredness
- Neurological symptoms: visual impairment/diplopia, headaches, carpal tunnel

Focused examination

- Hands
  - large hand span (spade-like), doughy skin, Phalen's/Tinel's test, increased palmar sweating, scar from carpal tunnel decompression
- Face
  - large nose and ears, look for prognathism (overbite) and malocclusion of the teeth, check tongue size
- Eyes
  - visual fields (bitemporal hemianopia)
- Extras
  - ask for blood glucose measurement and blood pressure

Questions

1 'What are the causes of acromegaly?'
  - Pituitary adenoma secreting growth hormone is the commonest cause.
  - Other tumours secreting excess GNRH: adrenocortical, pancreatic islet cell, lung.

2 'What investigations are required to make the diagnosis?'
   - Oral glucose tolerance test: failure of an oral glucose load (75 g) to suppress GH secretion two hours after administration, thus GH levels remain elevated.
   - IGF-1 levels, used less frequently: elevated in parallel to GH levels.
   - Full pituitary hormone screen: T4, TSH, LH, FSH, testosterone/oestradiol, cortisol, prolactin.
   - MRI brain/pituitary fossa: identify size and location of tumour.
   - Full visual field checks.

3 'What is the cause of the bitemporal hemianopia?'
   - Compression of the optic nerve at the optic chiasm from a mass arising from the underlying pituitary fossa.

4 'What is the treatment for acromegaly?'
   - Surgery
     - mainstay of treatment; transsphenoidal hypophysectomy for smaller tumours and a transfrontal approach for larger tumours
   - Medical therapy
     - may be needed before and even after surgery to reduce tumour bulk; agents include somatostatin analogues (octreotide and lanreotide), or dopamine agonists such as bromocriptine and cabergoline
   - Radiotherapy
     - reserved for patients unfit for surgery as very slow action on reducing tumour size

Patient welfare and concerns

- Cosmetic: regarding facial appearance and large stature/hands/feet.
- Underlying cause: Is this malignant? Concerns regarding a 'brain tumour'.
- Treatment: Is it curative?
- Complications: diabetes, hypertension and prognosis.

Candidate expectations

- Recognise the range of symptoms pointing towards acromegaly.
- Investigate possible unifying diagnoses for the collection of symptoms.
- Formulate the correct diagnosis and convey this to the patient.

- Examine for features consistent with acromegaly and any complications, e.g. visual field defects.
- Explain available treatment strategies with reiteration that this is not a malignant tumour.

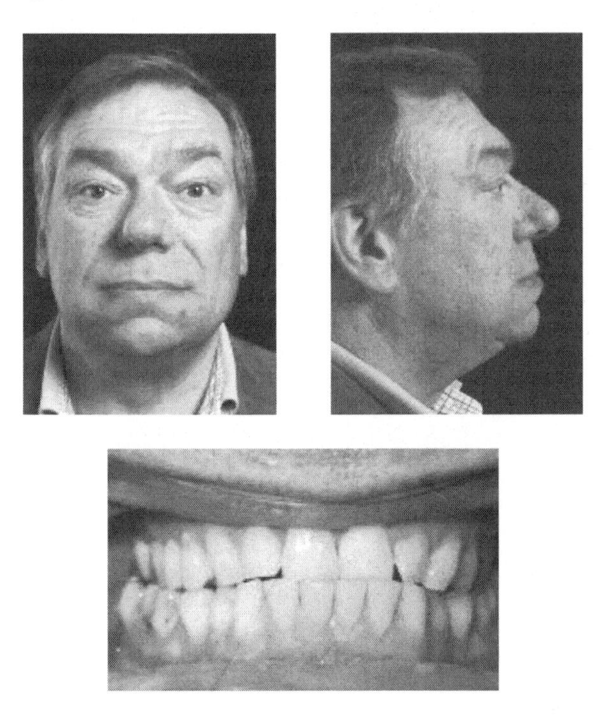

Figures 5.1, 5.2, 5.3 Facial features of acromegaly

Ankylosing spondylitis

'This patient has been experiencing episodes of backache. Please ask any relevant questions and proceed as appropriate.'

Focused history

- Duration nature and location of backache
- Age and gender of patient (males affected > females)
- Morning stiffness: duration
- Falls, trauma, injury to spine
- Deformity to spine
- Other joints affected: sacroiliac joints, hips, knees
- Family history of back pain/problems
- Neurological symptoms: ensure no bladder/bowel disturbance (all histories of back pain), paraesthesia/numbness/weakness in limbs
- Other features: chest pain, breathlessness, eye symptoms (pain, redness, floaters)
- Patient welfare/concerns: deformity, genetic link, mobility aids

Focused examination

- Spine
  - ask patient to stand up, back and front fully exposed (retain underwear to preserve dignity).
  - kyphotic spine, compensatory hyperextension of the neck ('question mark' posture)
  - reduced spinal movements: rigid, immobile spine
  - increased AP diameter of chest wall
- Cardiac
  - listen to aortic area and left sternal edge for early-diastolic murmur of aortic regurgitation
- Chest
  - fine apical fibrotic crepitations
- Eyes
  - iritis, visual acuity check
- Gait
  - likely antalgic; will make the spinal deformity more obvious

Questions

1 'What are the immunological associations with ankylosing spondylitis?'
  - Seronegative spondyloarthropathy.
  - HLA-B27 positive in >90% of individuals.
  - TNF-α and IL-1 also implicated in disease activity.

2 'How is the diagnosis of ankylosing spondylitis made?'
  - Mainly from history and clinical examination with supporting radiological evidence.
  - Young patients (<40), possible family history.
  - Plain radiograph: erosions and fibrosis/sclerosis of the sacroiliac joints, squaring of the vertebra ('bamboo spine') due to ossification of the anterior longitudinal ligament and intervertebral spaces.
  - Blood tests: raised ESR/CRP during active inflammation, normocytic anaemia.

3 'What is the treatment for ankylosing spondylitis?'
  - No known cure: mainly symptomatic.
  - Conservative: encourage increased exercise, physiotherapy, exercises for maintaining good posture.
  - Medical therapy: NSAIDs for pain, DMARDs for immunomodulation.
  - Biological agents: TNF-α antagonists – infliximab, adalimumab, etanercept (help slow disease progression).
  - Surgery: very rarely spinal osteotomies can be performed to correct the deformity.

4 'What are the complications of ankylosing spondylitis?'
  - Respiratory: restrictive lung defect/reduced lung capacity due to restricted chest wall movement, apical lung fibrosis.
  - Cardiac: chronic aortitis leading to aortic regurgitation, conduction defects, cardiomyopathy.
  - Neurological: atlantoaxial instability/dislocation, sciatic, cauda equina syndrome.
  - Eyes: iritis, cataracts.
  - Amyloidosis (secondary/AA): hepatic and renal involvement.

Patient welfare and concerns

- Adequate management of pain.
- Concern regarding the impact upon activities of daily living; will this stop the patient working?

- Cosmetic concerns regarding possible spinal deformity.
- 'Is this inherited and could I pass this on to my children doctor?'
- Is there a cure?

Candidate expectations

- Recognise an inflammatory arthropathy, in particular ankylosing spondylitis, from a detailed history.
- Be aware of the need to rule out 'red-flag symptoms'.
- Understand the extra-articular manifestations of inflammatory arthropathies.
- Examine for features of ankylosing spondylitis including extra-articular manifestations, e.g. auscultate for aortic regurgitation and lung fibrosis.
- Explain that treatment aims are to maintain a good quality of life and prevent disease progression and complications.

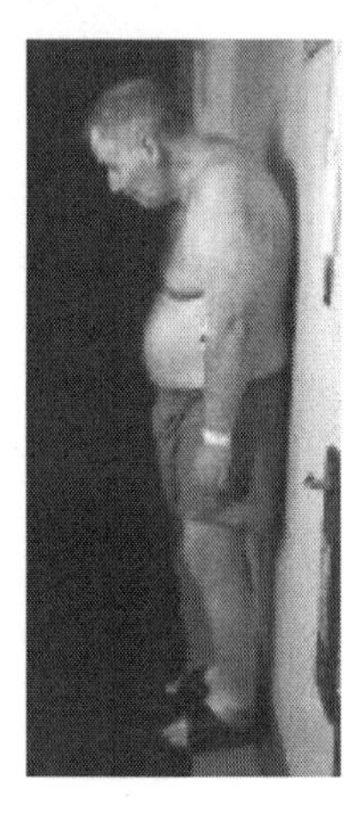

Figure 5.4 Characteristic stooped posture of ankylosing spondylitis

Chronic diarrhoea (coeliac disease)

'This patient has recently been diagnosed with irritable bowel syndrome and is particularly troubled by loose stools. They want to know what treatments are available to help. Please ask any relevant history to address the patient's concerns and reassure them as appropriate.'

Focused history

- When was the IBS diagnosed and by whom (GP/gastroenterologist)? What investigations have been done previously?
- How long has the patient been troubled with symptoms? Review the diagnosis and symptoms of diarrhoea (consistency of stools, frequency, nocturnal, blood PR/mixed with stools), abdominal pain/bloating, wind and nausea and vomiting.
- Precipitating factors: stress, particular foods (dairy products/wheat/gluten).
- Any sinister symptoms: weight loss, change in appetite, blood/mucus per rectum.
- Any other symptoms: rashes/skin changes, joint involvement, mouth ulcers, fistulae/abscesses.
- Rule out other causes: infective, thyroid disease, coeliac disease, lactose intolerance, pancreatitis, diabetes, carcinoid, iatrogenic (medication associated).
- Family history of inflammatory bowel disease.

Focused examination

- General
  - cachectic/malnourished, pallor, angular stomatitis
- Pulse
  - tachycardia (thyroid disease)
- Face/neck
  - pallor of conjunctiva/angular stomatitis/glossitis/mouth ulcers (evidence of malnutrition), obvious goitre

- Skin
  - inspect extensor surfaces and abdomen for skin rash: dermatitis herpetiformis (intensely itchy, vesicular, erythematous)

Questions

1 'What is the differential diagnosis of chronic diarrhoea?'
  - Inflammatory: ulcerative colitis, Crohn's, coeliac disease, chronic pancreatitis
  - Irritable bowel syndrome, lactose intolerance
  - Infective
  - Autoimmune: diabetes, thyrotoxicosis
  - Carcinoid syndrome
  - Iatrogenic: laxatives, antibiotics

2 'What is coeliac disease?'
  - Autoimmune disorder affecting the small intestine.
  - Tissue transglutaminase enzyme modifies gliadin (gluten protein in wheat), triggering the formation of anti-tTG antibodies that cause a mucosal inflammatory response.
  - Anti-tTG antibodies attract T-cells towards the small intestinal mucosal lining causing villous atrophy and malabsorption of minerals, nutrients and fat-soluble vitamins.
  - There is an association between coeliac disease and HLA DQ2 allele positivity.
  - Villous atrophy is reversible with strict adherence to a gluten-free diet.

3 'What tests are available to help diagnose coeliac disease?'
  - Blood tests: check for anaemia, B12/folate/ferritin, vitamin D and calcium levels, clotting profile (vitamin K deficiency).
  - Immunological tests: IgA anti-tTG antibodies, IgG anti-tTG antibodies, anti-endomysial antibodies, anti-gliadin antibodies.
  - Total IgA (often IgA deficiency coexists with coeliac disease).
  - Endoscopy with duodenal bulb biopsy: looking for increased numbers of lymphocytes within the intestinal mucosa, proliferation of the crypts of Lieberkühn or loss of villi.

4 'What is the association with malignancy in coeliac disease?'
  - Various GI: oropharyngeal, oesophageal, small-bowel adenocarcinoma.
  - Enteropathy associated T-cell lymphoma: T-cell mediated lymphoma affecting the small intestine. Histological features are similar to those of coeliac disease itself. Occurs in untreated disease and carries a poor prognosis.

Patient welfare and concerns

- Impact of diarrhoea on daily life and quality of life.
- Is there a serious underlying pathology/cause?
- What does a diagnosis of coeliac disease mean?
- What are the treatment options? 'Will I have to pay for the gluten-free products?'

Candidate expectations

- Obtain a detailed and thorough history of the diarrhoea and rule out a sinister underlying cause quickly.
- Be aware of the range of causes of chronic diarrhoea and how to exclude them by means of the history. Be prepared to question a patient's diagnosis where appropriate.
- Check for features suggestive of a malabsorption syndrome and dermatological associations of coeliac disease.
- Stress the importance of a strict gluten-free diet to maintain well-being.

Osteoporosis secondary to Cushing's syndrome

'This patient has recently been diagnosed with diabetes mellitus and started on metformin tablets. She has a past history of chronic lower back pain for which she takes several painkillers. She has presented today with worsening backache and wants to know what more could be done about it.'

Focused history

- Is the diabetes a red-herring or it is pertinent to the history?
- Duration of current episode of back pain: onset (acute/sudden or gradual worsening), severity (pain-score), character, site, radiation. Has the pain ever been this severe in the past? Analgesic history.
- Precipitating factors: heavy lifting/fall/trauma.
- Neurological symptoms: paraesthesia/weakness or paralysis in lower limbs, bladder or bowel dysfunction.
- Red-flag symptoms: weight loss, nocturnal pain, fevers.
- Relevant other history: risk factors for osteoporosis (steroid use, immobility, family history), hypertension, recent diagnosis of diabetes (possibly a coincidence, or is there underlying Cushing's syndrome?)

Focused examination

- General
  - examine patient standing: obvious spinal deformity, skin/palmar pigmentation if pigmented appearance, central obesity, round face, thin skin, bruising
- Spine
  - palpate vertebrae to localise symptoms; assess spinal movements as far as able
- Limbs
  - focused assessment of power in lower limbs, sensation and gait
- Extras
  - state you would examine perineum for perianal sensation and anal sphincter tone; ask for blood glucose measurement and blood pressure reading

Questions

1 'What are the WHO definitions of osteoporosis?'
- Diagnosis is based upon a T-score from bone mineral density (BMD) using dual-energy X-ray absorptiometry (DEXA) scanning. Bone density T-score ≤2.5 standard deviations below normal values for healthy young adults is defined as osteoporotic.
  - T-score greater than –1.0 = normal
  - T-score between –1.0 and –2.5 = osteopenia
  - T-score lower than –2.5 = osteoporosis
  - T-score < –2.5 with one or more fractures = severe osteoporosis

2 'What are the risk factors for developing osteoporosis?'
- Family history
- Disease associations: Cushing's syndrome, malabsorption, hyperparathyroidism, chronic inflammatory arthropathy
- Toxins: excess alcohol or caffeine intake, smoking
- Prolonged immobility or inactivity
- Underweight
- Early menopause, late menarche, post menopause, bilateral oophorectomy, hypogonadism in males
- Drugs: prolonged steroid use (>7.5 mg prednisolone daily for six months), prolonged use of low molecular weight heparin

3 'What strategies can be used in the treatment of osteoporosis?'
- Prevent falls: appropriate footwear, OT/physio assessments (home adjustments), avoid sedative medications
- Reduce impact of falls: hip protectors
- Drug therapies
  - reduce bone resorption: bisphosphonates (zoledronic acid, ibandronic acid, alendronic acid), raloxifene (selective oestrogen reuptake inhibitor), calcitonin
  - aid new bone formation: calcium and vitamin D supplements
  - reduce bone resorption and increase new bone formation: strontium ranelate

4 'What is the difference between Cushing's syndrome and disease?'
- Syndrome: due to cortisol excess which can be endogenous or exogenous
  - endogenous: adrenal gland tumours/benign adrenal gland hyperplasia, ectopic secretion of cortisol or ACTH from tumours elsewhere

    - exogenous: glucocorticoids, or tumours secreting excess cortisol/ACTH.
  - Disease: due to pituitary adenoma producing excess ACTH
5 'What are the features of cortisol excess?'
  - Hypertension
  - Diabetes mellitus
  - Osteoporosis
  - Infertility, impotence
  - Electrolyte disturbance: hypokalaemia, hypernatraemia, hypercalcaemia
  - Depression/low mood
  - Increased susceptibility to infections

Patient welfare and concerns

- Concerns regarding the chronic back pain and the need for a diagnosis.
- The impact of the back pain on daily life and ability to work.
- What is Cushing's syndrome and how is this related to the back pain?
- Is there a treatment: medical or surgical, other than simple analgesia, for the back pain/osteoporosis?

Candidate expectations

- Obtain a thorough history of the back pain and quickly rule out 'red-flag signs'.
- Rule out spinal cord compression in all patients presenting with worsening back pain.
- Search for causes of osteoporosis if the patient has had an osteoporotic compression fracture leading to the back pain (in this case possible Cushing's).
- Thoroughly examine any deformity of the spine and the range of spinal movements. Check for any neurological deficit and offer to check for perianal sensation and tone.
- Explain the likely diagnosis.
- Explain that bisphosphonates are first-line agents in the management of osteoporosis, both for treatment and prevention of further fractures.

Diabetic retinopathy

'This patient has noticed that his vision has become blurred at times. Please ask any relevant questions and proceed as appropriate.'

Focused history

- Duration of symptoms: progressive change rather than sudden onset
- Any visual loss
- Assess diabetic control
- Other microvascular complications: nephropathy, neuropathy
- Macrovascular complications: peripheral vascular disease, ischaemic heart disease, stroke

Focused examination

- Eyes
  - glasses, reactive pupils; check visual acuity
- Fundoscopy
  - cataracts, features of diabetic retinopathy (non-proliferative, proliferative, maculopathy, photocoagulation scars)
- Extras
  - evidence of neuropathy
  - ask for blood glucose measurement and urine dipstick for protein

Questions

1. 'What are the features of non-proliferative diabetic retinopathy?'
   - Microaneurysms (dot haemorrhages)
   - Blot haemorrhages
   - Hard exudates
   - Soft exudates (cotton-wool spots)
2. 'What are the features of proliferative diabetic retinopathy?'
   - Features of non-proliferative diabetic retinopathy, plus evidence of new vessel formation.
   - Photocoagulation scars are evidence of treatment.
3. 'What are the features of diabetic maculopathy?'
   - Any features of diabetic retinopathy at or near the macula.
   - Most commonly there is circinate formation of hard exudates.

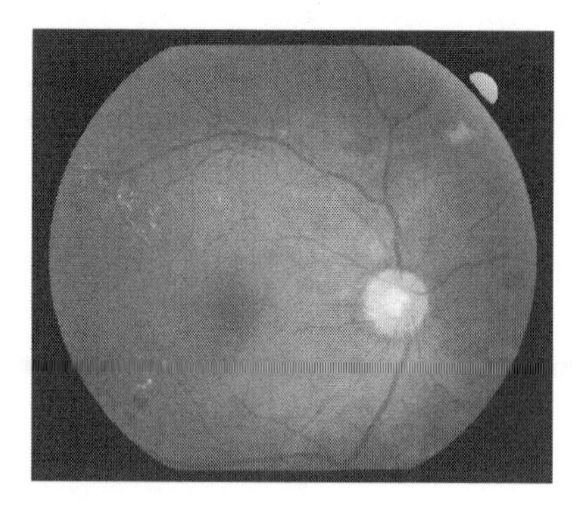

Figure 5.5 Retina displaying hard exudates, dot and blot haemorrhages, and a soft exudate (at 1 o'clock)

4 'What effect does glycaemic control have on the risk of diabetic retinopathy?'
   — This was assessed by the UK Prospective Diabetes Study (UKPDS). The study showed that for every 1% decrease in Hb A_{1c}, there was a 35% risk reduction in microvascular complications.

Patient welfare and concerns

- Vision: concerns of visual loss
- Prevention: prevention of further deterioration
- Treatment: Is treatment possible?

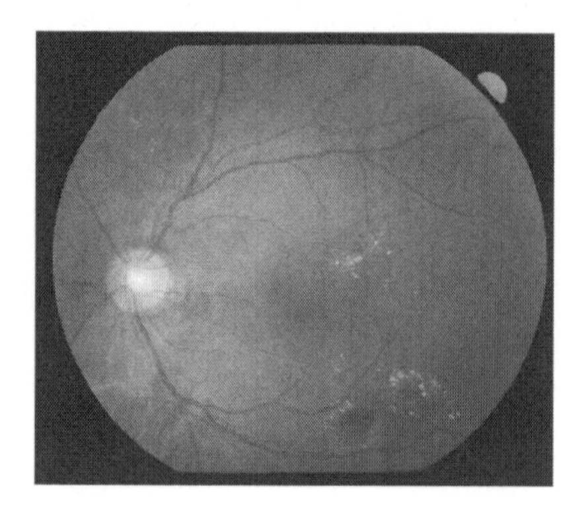

Figure 5.6 Hard exudates near the macula

Candidate expectations

- Recognise that diabetic retinopathy can be asymptomatic or can present with visual disturbance.
- Recognise the features of diabetic retinopathy.
- Understand the stages of diabetic retinopathy.
- Formulate the correct diagnosis and convey this to the patient.
- Generate a management plan, recognising when urgent intervention is necessary to save vision.

Facial rash

'This patient has noticed a rash, mainly over his face. He thinks that his skin burns easily, and first noted this after playing golf two weeks ago. He wants to know what has caused it and would like a solution if possible.'

Focused history

- Duration of symptoms
- Areas affected (and note distribution over face)
- Exacerbating features
- Features of rash: may be itchy, no ulceration
- Potential causes
  - arthritis, oral ulceration, breathlessness, neurological features (SLE)
  - thyroid disease, breathlessness, visual change/'night glare' (amiodarone)
  - dark urine, abdominal pain, confusion (porphyria)

Focused examination (if on amiodarone)

- Rash
  - photosensitive rash all over the face
  - may be excoriations from itching
  - often also appears on back of hands, ears and superior part of chest
- Extra features
  - pacemaker
  - thyroid mass
  - apical lung fibrosis
- As well as photosensitivity, prolonged use of amiodarone can cause a slate-grey/violaceous discolouration of exposed skin.

Questions

1 'What drugs cause photosensitivity?'
  - Amiodarone
  - OCP
  - Thiazides

— Sulphonamides
— Phenothiazines

2 'What are the other causes of a photosensitivity rash?'
— SLE/DLE
— Porphyria
— Pellagra
— Rosacea

3 'What treatment options are available for amiodarone photosensitivity?'
— Discontinuation of amiodarone
— Barrier methods (sun-block, hats and long-sleeved clothing)

Patient welfare and concerns

- Cosmetic concerns regarding skin changes.
- Resolution of the rash (photosensitivity and dusky-grey skin pigmentation usually resolve, though this may take up to two years after stopping the drug).
- The need to stop amiodarone; what are the alternatives?
- Apart from stopping the drug, what else can be done?

Candidate expectations

- Obtain a history of the rash.
- Recognise that the patient has a photosensitive rash.
- Realise that the cause is amiodarone, after the patient volunteers that they are on this drug.
- Examine the face, but also other systems affected by amiodarone.
- Explain the likely diagnosis.
- Explain possible treatment options, including alternative antiarrhythmics where appropriate.

Facial nerve palsy

'This patient has been experiencing difficulty with closing their eyelids on one side. As a result the eye is becoming dry and irritable. They are anxious to know the cause of the symptoms and what can be done. Please ask any relevant questions and proceed as appropriate.'

Focused history

- Which eyelid is affected? What appears to be the problem? Does the eyelid shut at all and if so how much?
- Duration of symptoms?
- Are there any associated problems: speech, swallowing, taste or hearing impairment (tinnitus*, hyperacusis or deafness*)
- Any vertigo*/nausea*/vomiting or abnormality in gait* noted
- Any associated facial droop or distorted angle of mouth, change in taste
- Any recent viral (herpes) infection, surgery to the neck (parotid gland), inner ear or mastoid
- Medical history/family history: neurofibromatosis +/– acoustic neuroma with or without previous surgery

*Be alert to these symptoms as may be indicative of acoustic neuroma

Focused examination

- General
  - Examine the patient at rest to assess facial symmetry
  - Any obvious drooping of the angle of the mouth, ptosis (unilateral); ask patient to close their eyes; assess ability to close fully; is there a tarsorrhaphy scar?
- Movements
  - Assess muscles supplied by facial nerve (demonstrate the movements yourself)
  - Ask patient to squeeze their eyes tightly shut, raise their eyebrows, blow out their cheeks, smile, show you their teeth
- Extras
  - Look behind the ear for a mastoid surgery scar; examine for any cranial scars from acoustic neuroma removal

- Assess for any hearing disturbance (cranial nerve VIII involvement)
- Assess gait (cerebellar/ataxic)

Questions

1 'What are the causes of unilateral CN VII palsy?'
- UMN lesion
  - cerebellopontine angle (CNs V, VI, VII, VIII and loss of taste on anterior 2/3 of the tongue): acoustic neuroma, meningioma
  - pontine lesion: demyelination (MS), vascular lesion
- LMN lesion
  - Bell's palsy: commonest cause; caused by herpes simplex type 1
  - Ramsay Hunt's syndrome; caused by herpes zoster virus
  - parotid gland tumour/surgery
  - facial neuroma
  - cholesteatoma
  - mononeuritis multiplex (diabetes, SLE, PAN, sarcoid, amyloid, Wegener's granulomatosis)
  - trauma

2 'What are the causes of bilateral LMN cranial nerve VII palsy?'
- MND
- Guillain–Barré syndrome
- Bilateral Bell's palsy
- Lyme disease
- Myasthenia gravis
- Sarcoidosis
- Moebius' syndrome (inherited rare form due to underdevelopment of cranial nerves VI and VII)

3 'Which condition is associated with an acoustic neuroma?'
- Neurofibromatosis type 2: defect on chromosome 22q12; this condition results in bilateral acoustic neuromas

4 'What is the management of an acoustic neuroma?'
- Conservative: observe the tumour size and growth
- Radiotherapy: to retard tumour growth
- Surgical resection

Patient welfare and concerns

- Aesthetic: facial asymmetry secondary to the nerve palsy due to medical/surgical cause

- Will the facial nerve palsy resolve?
- Is there any treatment for this?
- Will the deafness resolve or will it be permanent? Is there any treatment for this?

Candidate expectations

- Obtain a thorough history of symptoms of a CN VII lesion.
- Be aware of the range of causes leading to CN VII palsy.
- Examine carefully to differentiate between UMN and LMN CN VII palsy.
- Provide the patient with a clear explanation of the problem and the likely effects, e.g. if it is a Bell's palsy, there is a good chance of recovery; however, if it is a surgical consequence then that is unlikely to be the case.
- Explain that any deafness is likely to be permanent and offer treatment solutions if possible.

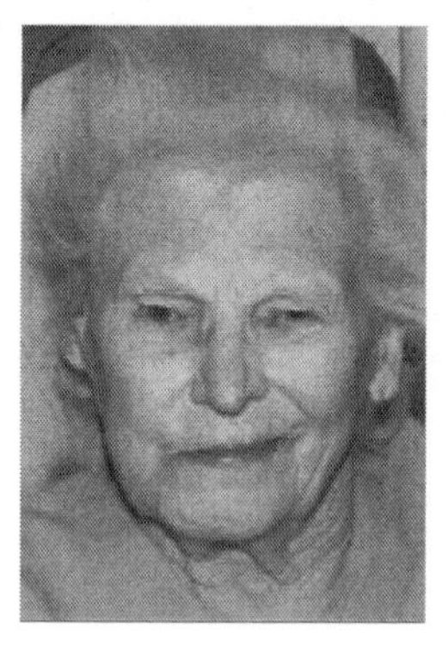

Figure 5.7 Facial nerve palsy

Headache (idiopathic intracranial hypertension)

'This 30-year-old lady with a family history of diabetes mellitus had a glucose tolerance test six months ago and was found to have impaired fasting glycaemia. Since then she has been complaining of headaches and visual disturbance, getting much worse in the last two weeks. She does not usually wear glasses and wants to know what can be done to improve the symptoms.'

Focused history

- Nature of visual disturbance: diplopia or blurred vision, floaters/ flashing lights, visual loss
- Duration of visual disturbance: stable or deteriorating
- Effect on visual acuity: any recent eye tests
- Headaches: worse in morning, on bending/coughing/sneezing; vomiting, nausea
- Associated symptoms: paraesthesia, weakness, hearing disturbance
- Relevant other history: weight loss/gain, obstructive sleep apnoea, drug history (vitamin A derivatives, tetracyclines, oral contraceptive pill)

Focused examination

- General
  — obese, any obvious cranial nerve palsies
- Eyes
  — enlarged blind spot, ophthalmoplegia (abducens nerve palsy)
- Fundoscopy
  — assess for papilloedema and optic atrophy
- Extras
  — check for scars from VP shunts; ask for blood glucose measurement and blood pressure

Questions

1 'What are the causes of papilloedema?'
  — Space-occupying lesion (tumour/abscess)

- Idiopathic intracranial hypertension (IIH)
- Hypertensive encephalopathy
- Infection: encephalitis, meningitis
- Vascular: intra/extra-axial haemorrhage, venous sinus thrombosis
- Drugs: tetracyclines, vitamin A derivatives

2 'What are the fundoscopic features of papilloedema?'
- Disc hyperaemia, blurred margins, absent venous pulsation
- Elevation of the disc with obscured vessels at the disc margin
- Loss of the cup with obscured vessels in the disc
- Bulging disc with all vessels obscured (this will eventually lead to optic atrophy)

3 'What are the causes of optic atrophy?'
- Congenital: Friedreich's ataxia, Leber's hereditary optic neuropathy
- Acquired
  - vascular: ischaemic (including temporal arteritis)
  - inflammatory: multiple sclerosis, Devic's disease
  - compression: optic nerve tumour, Graves's ophthalmopathy, glaucoma
  - IIH (untreated)
  - nutritional deficiencies: vitamin B12, folate
  - toxins: tobacco, alcohol, ethambutol, ethylene glycol, lead, cyanide, carbon monoxide
  - infective: syphilis

4 'What treatments can be used to treat IIH?'
- Medical
  - weight loss if obese
  - stop any contributing medications
  - diuretics, e.g. acetazolamide
  - steroids: beneficial in inflammatory conditions/causes
  - repeated lumbar punctures
- Surgical
  - optic nerve sheath decompression and fenestration
  - lumboperitoneal/ventriculoperitoneal shunt

Patient welfare and concerns

- Concerns regarding the visual disturbance: Could I lose my sight?
- Concerns regarding the associated headache: Is this a brain tumour?

- How can you be certain that this is a benign condition and that there is not sinister underlying pathology?
- What are the treatment options? Would having a shunt inserted require brain surgery?

Candidate expectations

- Obtain a thorough history regarding the visual disturbances and headache.
- Identify features suggestive of raised intracranial pressure and/or a space-occupying lesion, e.g. weakness/paralysis, seizures.
- Carefully assess visual acuity, visual fields and ocular fundi.
- Explain the likely diagnosis of IIH but make patient aware of the need to exclude a space-occupying lesion.
- Explain that treatment strategies involve both medical and surgical options.

Hyperthyroidism

'This patient has a tremor and weight loss. Please ask any relevant questions and proceed as appropriate.'

Initial impressions

- Exophthalmos, a peripheral tremor, goitre (or scar over the neck) and general sweatiness point towards a diagnosis of hyperthyroidism.

Focused history

- Symptoms of thyroid disease
  — neck lump, 'bulging eyes' (exophthalmos)
- Features of thyrotoxicosis
  — weight loss and increased appetite, heat intolerance, sweating, diarrhoea, oligomenorrhoea, emotional lability and irritability
- Treatment instituted
  — beta-blockers, carbimazole/propylthiouracil

Focused examination

- Hands
  — acropachy, sweaty palms, tachycardia/arrhythmia, fine tremor
- Face
  — exophthalmos and proptosis, lid retraction, lid lag, ophthalmoplegia
- Neck
  — thyroidectomy scar, goitre (diffuse or nodular), thyroid bruit
- Legs
  — pretibial myxoedema
- Extras
  — evidence of other autoimmune disease

Questions

1 'What are the causes of hyperthyroidism?'
  — Graves's disease
  — Toxic multinodular goitre
  — Thyroid adenoma (toxic)
  — De Quervain's thyroiditis

— Postpartum thyroiditis
— Drugs (lithium, amiodarone)

2 'What are the causes of a goitre?'
— Nodular
  - multinodular goitre
  - adenoma
  - carcinoma
— Diffuse
  - 'simple' (physiological)
  - Graves's disease
  - Hashimoto's thyroiditis
  - De Quervain's thyroiditis

3 'What signs of thyroid disease are specific to Graves's disease?'
— Exophthalmos
— Ophthalmoplegia
— Pretibial myxoedema

4 What are the 'hyperthyroid emergencies'?
— Thyroid storm
— Exophthalmos causing fixed gaze, diplopia or decreased acuity (may lead to optic nerve compression)
— Cardiac failure (high output)

Patient welfare and concerns

- What treatment options are available?
- Is my condition curable?
- Is my condition life threatening?

Candidate expectations

- Recognise early that the patient has thyroid disease.
- Assess the patient's thyroid status (hyper/hypo/euthyroid).
- Recognise signs specific to Graves's disease.
- Produce a list of causes for the thyroid disturbance specific to the patient.

Hypothyroidism

'This patient presents complaining of feeling tired all the time. They have also noticed a neck lump and are concerned as to what could be the cause of the symptoms. Please ask any relevant questions and proceed as you feel appropriate.'

Focused history

- What is the most concerning problem: tiredness or neck mass/lump?
- When did the neck lump first appear? Is it enlarging? Site of neck lump (central/peripheral)? Any other lumps elsewhere?
- Symptoms of thyroid disease: over/underactive?
  - weight loss/gain, appetite
  - intolerance to heat/cold
  - mood instability, irritability
  - bowels: diarrhoea/constipation
  - palpitations
  - tiredness: is it due to disease or associated/other comorbidity (anaemia, adrenal insufficiency, malignancy)?
  - impaired memory/cognition
  - eyes: ophthalmoplegia, protrusion
  - dry skin, hair thinning/loss, menstrual irregularities
- Sinister symptoms of neck mass: breathlessness, hoarse voice, dysphagia, weight loss, night sweats
- Family history of thyroid disease and malignancy
- History of autoimmune disease: diabetes mellitus, rheumatoid arthritis, adrenal insufficiency/Cushing's disease

Focused examination

- General
  - Hypothyroid facies (dry coarse skin, hair thinning/loss); inspect neck mass from front and side; ask patient to take a sip of water, hold it, and then swallow, to see if mass moves with swallowing.

- Palpation
  - Palpate the mass from behind and feel how far it extends; ask the patient to take a sip of water; feel for whether the mass moves on swallowing; ask the patient to protrude their tongue; does it move with this action? Palpate for lymph nodes.
- Percuss
  - Retrosternal extension.
- Auscultate
  - Bruit over the thyroid.
- Extras
  - Pulse, pallor of conjunctiva, hoarseness of the voice.
  - Examine eye movements and check for any lid lag/retraction, chemosis and protrusion.
  - Check reflexes and look for any pretibial myxoedema.

Questions

1 'What are the causes of hypothyroidism?'
   - Thyroid: Hashimoto's thyroiditis, iodine deficiency, surgery, postpartum thyroiditis
   - Pituitary: tumour, radiotherapy, surgery
   - Hypothalamus (TRH deficiency)
   - Drugs: radioiodine therapy, lithium, amiodarone, overtreatment for hyperthyroid

2 'How do you diagnose hypothyroidism?'
   - TSH: raised (if primary thyroid problem)
   - Free T4: low
   - Free T3: low or low/normal
   - Thyroid antibodies: TPO

3 'What is compensated euthyroidism?'
   - Raised TSH levels above the normal range (but not as significantly as in hypothyroidism) in the presence of a normal T4
   - Risk of developing into overt hypothyroidism

4 What are the causes of a thyroid lump?
   - Solitary lump: thyroid cyst, adenoma, malignant disease
   - Diffuse enlargement: multinodular goitre, autoimmune thyroid disease (Graves's and Hashimoto's)

5 'What are the types of thyroid cancer?'
   - Papillary: most common
   - Follicular

- Medullary: arise from parafollicular calcitonin-secreting C cells
- Anaplastic: most aggressive
- Lymphoma

Patient welfare and concerns

- What is the cause: Is this a malignant mass?
- Cosmetic: obvious neck lump
- Associated features: hoarse voice, swallowing difficulties, symptomatic thyroid disease
- Treatment options

Candidate expectations

- Limit the diagnosis to thyroid over/underactivity to aid diagnosis via focused history.
- Thoroughly assess a thyroid mass and examine for any associated features of thyroid disease.
- Formulate the correct diagnosis and convey this to the patient.
- Inform the patient of the management plan. Reassure the patient that thyroid malignancy is just one of many causes of a thyroid mass, and is not the most common cause.

Proximal myopathy

'This lady has been recently seen by her doctor for generalised aches and pains present for several months. She is now complaining of weakness in her limbs. She is housebound due to severe osteoporosis and often does not go out for weeks at a time. Please ask any relevant questions and proceed as you feel appropriate.'

Focused history

- What is the nature of the weakness? Is it symmetrical and which limbs are affected? Is it proximal/distal/widespread?
- How long has the weakness been present? How rapidly has it progressed?
- Has there been any associated muscle wasting? Are the muscles tender?
- Has she developed any new back pain suggestive of a recent osteoporotic fracture?
- Ask what impact the symptoms have on daily life to help decipher the pattern of weakness:
  - proximal myopathy: difficulty getting up out of chairs, washing and dressing, combing hair
  - distal weakness: difficulties with fine movements/tasks (writing, doing up buttons).
- Which joints are involved? Any symptoms suggestive of synovitis?
- Sinister symptoms: weight loss, fevers, speech/swallowing impairment, bladder/bowel dysfunction.
- PMH/cause of osteoporosis: endocrine disease (Cushing's, thyroid disease), chronic steroid use, MND, vitamin D deficiency, immobility, alcohol excess, smoking, immobility.
- Drug history: prolonged steroid use, statin therapy, vitamin D supplements.

Focused examination

- General
  - Gait should be normal.
  - There should not be any evidence of fasciculation on inspection.
  - Observe any areas of muscle wasting/disuse atrophy.

  - Evidence of trauma/deformity.
- Palpation
  - Check that the muscles are not tender to palpation.
  - Assess muscle power throughout each limb; pay close attention for proximal muscle weakness, power will be reduced in affected areas; the signs will be symmetrical.
  - If a chair is available, ask the patient to sit on it and get up off it without using their hands.
- Extras
  - Examine the spine for any obvious deformity.
  - Examine the hands: rule out dermatomyositis, rheumatoid disease.
- The differential diagnosis here is wide. A detailed drug history is crucial to eliminate a drug cause, for example, statin-induced myopathy. In a housebound individual who has little sunlight exposure, never forget vitamin D deficiency as a cause of proximal myopathy.

Questions

1 'What are the manifestations of vitamin D deficiency?'
  - Rickets: impaired growth and deformity of long bones in children
  - Osteomalacia: reduced bone mineralisation resulting in proximal muscle weakness
  - Osteoporosis: reduced bone density and increased fracture risk

2 'What are the risk factors for vitamin D deficiency?'
  - Nutritional/poor dietary intake
  - Malabsorption syndromes
  - Reduced sunlight exposure
  - Darker skin pigmentation
  - Lack of vitamin D in breast milk
  - Renal/liver impairment
  - Enzyme-inducing medication (e.g. anticonvulsant medications)

3 'What are the biochemical features of vitamin D deficiency?'
  - Reduced serum vitamin D levels
  - Reduced serum calcium levels
  - Reduced serum phosphate levels
  - Raised ALP
  - Raised PTH (secondary hyperparathyroidism)

Patient welfare and concerns

- What is the cause of the weakness? Is this reversible?
- Will they lose their independent mobility?
- Seriousness of the pathology: Is there an underlying sinister cause?
- How long will the symptoms take to recover following initiation of treatment?

Candidate expectations

- Be aware of the range of differential diagnoses contributing to proximal muscle weakness.
- The key in this case is to take a focused history around the causes of a proximal muscle weakness to help rule in/out a diagnosis.
- Examine the important areas of each system that help confirm the diagnosis.
- Formulate the correct diagnosis and convey this to the patient.
- Reassure the patient that there are many causes of a proximal myopathy and most of them are readily treatable; however, the main part of the management will be aggressive rehabilitation.

Psoriasis

'This patient has attended complaining of a significant change in the appearance of her nails. She is becoming increasingly distressed by the problem, which seems to be worsening despite several treatments by the GP.'

Focused history

- Duration of symptoms
- Nails: colour, change in appearance, brittleness, extent of the nail involved, nail-bed involvement, number of nails involved, history of trauma, treatments tried by GP
- Change in shape of fingers/toes
- Other system involvement
  - skin changes: plaques/scales, pustules, sites of skin involvement, itchy/dry skin
  - joints: number of sites (hands, feet, spine, hips), swelling, tenderness, deformity
- Family history of psoriasis/arthritis

Focused examination

- Hands
  - nail changes: discolouration, pitting, onycholysis, hyperkeratosis of nail bed
  - dactylitis, arthritis mutilans
- Skin
  - inspect the back, abdomen, extensor and flexor surfaces, scalp and posterior aspect of ears
  - plaque psoriasis: shiny, scaly, white plaques
  - guttate: numerous widespread, erythematous small 'teardrop' lesions
  - pustular: widespread red pustules (often tender)
  - flexural: plaques/areas of inflammation at limited sites

Questions

1 'What are the causes of psoriasis?'
 - Possibly immune-mediated via T-cell stimulated cytokine release causing excess proliferation and production of skin cells within the dermis
 - Genetic susceptibility: MHC-antigen mediated
 - Poststreptococcal infection (especially guttate form)
 - Drug-related

2 'What treatments are used in the management of psoriasis?'
 - Moisturising creams
 - Topical treatments: coal tar, dithranol, steroids, vitamin D analogues (calcipotriol) and retinoids/vitamin A derivatives (acitretin)
 - Phototherapy: UVA
 - Photo-chemotherapy: psoralen and ultraviolet A (PUVA)
 - Medications
   - oral steroids
   - DMARDs: methotrexate, cyclosporin, azathioprine; newer agents include tacrolimus and mycophenolate mofetil
   - biological agents: anti-TNF therapy (e.g. adalimumab), and monoclonal antibodies (e.g. infliximab)

Patient welfare and concerns

- Cosmetic/aesthetic concerns regarding plaques, nail changes and joint involvement.
- Curative or symptomatic control only.
- Duration of treatment (particularly immunosuppression): short course versus lifelong.
- Number of medications needed to control/abate symptoms.
- Long-term effects and risks of newer biological agents: susceptibility to infection, risk of malignancy.

Candidate expectations

- Take a relevant history regarding the nails and skin lesions.
- Obtain a detailed history for arthritis.
- Differentiate quickly from the history whether this is likely to be an inflammatory or degenerative joint problem.
- Examine the psoriatic plaques, appropriately commenting on colour, location, size of area and distribution affected, using correct terminology.

- Explain the diagnosis of psoriasis to the patient and the possible association between psoriasis and arthropathy.
- Explain that there are both local and systemic treatments that can be used to control symptoms.

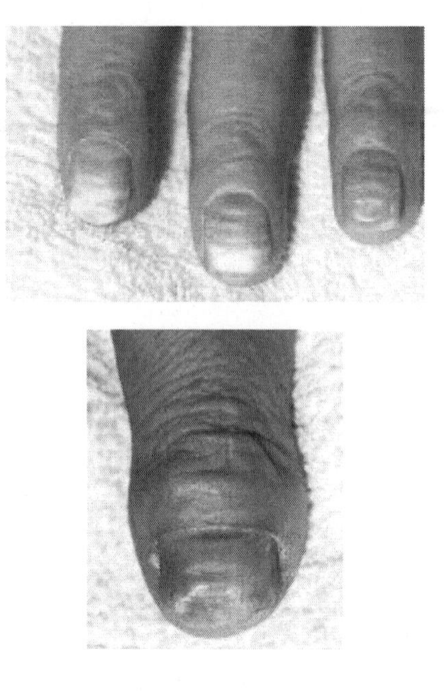

Figure 5.8 and 5.9 Nails showing psoriatic changes: onycholysis and pitting

Rheumatoid arthritis with carpal tunnel syndrome

'This patient has been complaining of a recent history of dropping objects. Ask any relevant questions and proceed as you feel appropriate.'

Focused history

- Hand dominance
- One or both hands affected
- Duration of symptoms: worsening/stable
- History of trauma or injury to hand/arm/shoulder
- Associated pain/numbness/paraesthesiae (including distribution)
- Joint synovitis, muscle wasting
- Extent of weakness: functional limitations, impact on QOL/ADLs
- Associated symptoms: fever, weight loss, other joint/system involvement
- Drug history

Focused examination

- General
  - evidence of rheumatoid arthritis, acromegaly, thyroid disease or pregnancy
- Hands
  - symmetrical arthropathy
  - thenar eminence wasting
  - deformity: ulnar deviation, MCPJ subluxation, swan-neck, boutonnière, Z-thumb
  - evidence of scars from previous tendon release or nerve decompression surgery
- Palpation
  - hand, wrist and elbow joints: active synovitis?
- Movements
  - test specifically the lumbricals, opponens pollicis, abductor pollicis brevis and flexor pollicis brevis
  - assess function: doing up a button, key grip, opening a door handle

- Neurological
  - Phalen's and Tinel's test to assess the median nerve

Questions

1 'What are the diagnostic criteria for rheumatoid arthritis?'
  - American College of Rheumatologists' criteria
    - morning stiffness lasting at least one hour on most mornings
    - arthritis affecting at least three or more joints
    - arthritis affecting the hands
    - symmetrical arthritis
    - subcutaneous nodules
    - positive rheumatoid factor
    - radiological evidence of rheumatoid arthritis on hand and wrist radiographs, features including erosions

2 'What are anti-cyclic citrullinated peptide (anti-CCP) antibodies?'
  - Antibody markers used for diagnosis of rheumatoid arthritis
  - Aid assessment of likely future progression to RA, in undifferentiated arthritis
  - More sensitive (~70%–75%) and specific (~95%–98%) than IgM rheumatoid factor antibodies
  - Prognostic value as marker of erosive disease

3 'What biological therapies are available for the treatment of rheumatoid arthritis?'
  - TNF-α blockers: infliximab, adalimumab, etanercept
  - Interleukin-1 (IL-1) blockers: anakinra
  - Monoclonal B-cell (anti-CD 20) antibodies: rituximab
  - Newer therapies: IL-6 blocker, tocilizumab

4 'What are the serious side effects associated with the use of biological therapies?'
  - Opportunistic infections: fungal, bacterial, viral (e.g. severe infection with varicella-zoster virus)
  - Activation of latent TB +/– progression to miliary TB
  - Anaphylaxis
  - Development of lymphoma
  - Hepatic damage
  - Central nervous system demyelinating disorders
  - Congestive cardiac failure

Patient welfare and concerns

- Concerns regarding the impact of an inflammatory arthropathy on daily life and the possible cosmetic effects of deformities.
- Concerns regarding systemic extra-articular manifestations.
- Impact of DMARDs/biological agents and their side effects.
- Risks to offspring in terms of developing rheumatoid arthritis in the future.

Candidate expectations

- Obtain a thorough history regarding the extent of joint involvement and duration of symptoms that clearly points to a diagnosis of inflammatory arthritis.
- Ascertain the impact of symptoms on daily life; can the patient still function as normal?
- Thoroughly examine all affected joints; assess active synovitis and identify any associated extra-articular manifestations.
- Explain the likely diagnosis of rheumatoid arthritis.
- Explain that treatment strategies range from anti-inflammatory agents to biological agents, though there are criteria for initiation of more aggressive treatments (biological therapies are not the first-line treatment).

Systemic sclerosis

'This patient has been complaining of painful fingers. Please ask any relevant questions and proceed as appropriate.'

Focused history

- Duration of symptoms, when they occur, elicit features of Raynaud's phenomena: colour changes in fingers, pain on exposure to the cold, nail changes
- Changes in appearance of the skin: thick, tight, shiny skin, oedematous fingers
- Any exacerbating/relieving factors: any treatments tried
- Extra-cutaneous features: reflux, dysphagia, breathlessness

Focused examination

- Hands
  - sclerodactyly (thick, shiny, tight skin), calcinosis, evidence of Raynaud's, nail changes/atrophy, digital infarcts; assess function
- Face
  - telangiectasia, microstomia
- Lungs
  - basal pulmonary fibrosis
- Extras
  - ask for blood pressure and urine dip (renal failure)

Questions

1 'What are the types of scleroderma?'
   - Localised scleroderma
     - localised skin involvement
   - Limited systemic sclerosis (also known as CREST syndrome)
     - calcinosis, Raynaud's phenomena, oesophageal dysmotility, sclerodactyly, telangiectasia
   - Diffuse systemic sclerosis
     - involves major organs (heart, kidneys and lungs) as well as skin; rapidly progressive

2 'Which other organs are affected?'
   - Kidneys: hypertension, chronic kidney disease progressing to ESRF

- Lungs: pleural effusions, pulmonary hypertension, interstitial pneumonitis/pulmonary fibrosis
- Cardiac: pericardial effusions/constriction, arrhythmias

3 'What are the treatment options?'
- Raynaud's: calcium-channel blockers, e.g. nifedipine, ACE inhibitors
- Reflux/oesophageal dysmotility: proton pump inhibitors, H2-antagonists
- Hypertension/renal protection: ACE inhibitors
- Skin/joint/lung: immune modulation: steroids, steroid-sparing agents, cyclophosphamide

4 'What immune tests can be used to help identify the cause?'
- Diffuse systemic sclerosis: anti-SCL 70 antibodies (against topoisomerase I)
- Limited systemic sclerosis: anti-centromere antibodies
- U1-RNP: indicative of a mixed connective tissue disorder (systemic sclerosis, SLE, polymyositis)

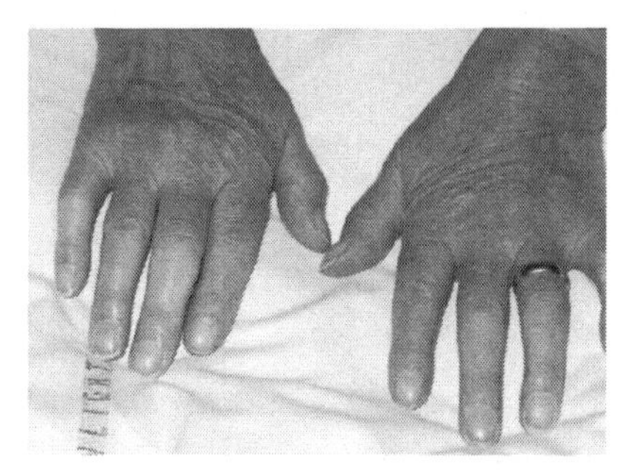

Figure 5.10 Sclerodactyly as part of systemic sclerosis

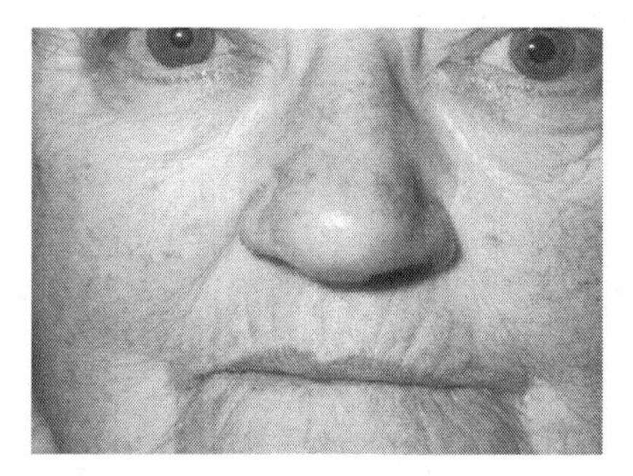

Figure 5.11 Facial features of systemic sclerosis: telangiectasia, microstomia

Patient welfare and concerns

- Concerns regarding skin changes in hands in particular relating to the weather.
- Is the reflux an unrelated separate issue to the Raynaud's/skin features?
- What are the other complications of the condition?
- What is the prognosis of the condition if there are cardiac, renal or pulmonary manifestations?

Candidate expectations

- Obtain a thorough history of the skin features.
- Recognise that the patient has Raynaud's and look for an underlying cause of this.
- Realise that the cause is systemic sclerosis.
- Examine the hands and face and other areas affected by systemic sclerosis.
- Explain the likely diagnosis.
- Explain the range of treatment options and the importance of preventing disease progression or relapse.

Station 5 summary

- It is crucial to dedicate substantial revision time to station 5.
- Be sure to look out early for 'spot diagnoses'.
- As well as looking for facial and hand features in acromegaly, it is vital to assess visual fields as this gives some clue towards the pathology (micro/macroadenoma).
- In a case of back pain (such as ankylosing spondylitis), be sure to rule out spinal cord compression/cauda equina.
- When reading the lead-in to a case, be prepared to question a previously made diagnosis if the evidence points against this; features of malabsorption are not consistent with irritable bowel syndrome.
- In cases of diabetes always consider other micro- and macrovascular complications.
- Be able to assess a patient's thyroid status clinically.
- In a case of psoriasis always look for evidence of arthritis.
- In any joint disorder it is crucial to assess function.
- Take note of any patient concerns throughout and attempt to resolve them later in the consultation.

Hawks and doves

As soon as you start revising for PACES, you should ask Registrars and Consultants to take you round to watch you examine patients, ask you questions and critique your performance. Very quickly you'll realise that some of these Registrars and Consultants are Hawks and some are Doves.

What is a Hawk? What is Hawkish behaviour? Hawks are tough! They believe in tough love, they give you a hard time, they expect high standards, they knock you off your perch and they keep you honest. Hawks are negative and pessimistic. They pick you up on every omission or fault in your examinations, they ask you difficult questions and they critique you harshly.

Like Yin and Yang, Doves are the opposite. Dovelike in behaviour, they are gentle! They encourage, they flatter, they appreciate your efforts, they build your confidence, and they let you fly. Doves are positive and optimistic. They praise your thorough examinations, they ask sensible questions that you can answer, and they critique you fairly.

These stereotypes are not absolute but every registrar or consultant who takes you round will either have hawkish tendencies or be dovelike. But does this matter? Who cares? Does it change anything? The answer is yes! It really matters – I'll explain why.

During a typical PACES revision session in the first 2–3 weeks, you're learning your examinations, you're learning how to present, you're gaining in confidence but it's all quite daunting. The last thing you need is a Hawk to swoop down and tear you to shreds, destroying your confidence before you've even started. Go for the Doves, to break you in gently.

After 2–3 weeks, your examinations become slicker, your presentations are smoother and you feel increasingly confident. You start to think you can do it, you can be good enough and you can make it. But PACES is difficult, examiners are not always nice, they can be tough. Now what you need are Hawks to take you round, give you a kick up the backside, put the fear back into you, keep you honest and working hard, and not lazy or sloppy with your examinations and presentations.

Finally, the exam approaches, about a week away. The Hawks and Doves have worked well, building your confidence and keeping you honest. You start to peak, you're really good and have every chance of passing. But the exam is daunting, you become nervous and you need a strong will and a positive mental attitude to perform on the day. The last thing you need is a Hawk to tear you down before your exam, destroying you and making you think the whole PACES experience is all too much. Seek those reassuring Doves again!

Good luck.

Ajay M Verma

Index